T0139091

Computer Network Simulation Using NS2

Computer Network Simulation Using NS2

Ajit Kumar Nayak
Satyananda Champati Rai
Rajib Mall

CRC Press
Taylor & Francis Group
Boca Raton London New York

CRC Press is an imprint of the
Taylor & Francis Group, an **informa** business

CRC Press
Taylor & Francis Group
6000 Broken Sound Parkway NW, Suite 300
Boca Raton, FL 33487-2742

© 2016 by Taylor & Francis Group, LLC
CRC Press is an imprint of Taylor & Francis Group, an Informa business

No claim to original U.S. Government works

Printed on acid-free paper
Version Date: 20160421

International Standard Book Number-13: 978-1-4987-6854-2 (Hardback)

Library of Congress Cataloging-in-Publication Data

Names: Nayak, Ajit Kumar, author. | Rai, Satyananda Champati, author. | Mall, Rajib, author.
Title: Computer network simulations using NS2 / Ajit Kumar Nayak, Satyananda Champati Rai, and Rajib Mall.
Description: Boca Raton : Taylor & Francis, CRC Press, 2016. | Includes bibliographical references and index.
Identifiers: LCCN 2016003396 | ISBN 9781498768542 (alk. paper)
Subjects: LCSH: Computer networks--Computer simulation. | NS (Electronic resource)
Classification: LCC TK5105.5 .N393 2016 | DDC 004.60285/53--dc23
LC record available at https://lccn.loc.gov/2016003396

Visit the Taylor & Francis Web site at
http://www.taylorandfrancis.com

and the CRC Press Web site at
http://www.crcpress.com

Printed and bound in the United States of America by Publishers Graphics, LLC on sustainably sourced paper.

CONTENTS

PREFACE

This book is intended to help students understand certain practical aspects of computer networking. We focus on simulation of basic computer networking protocols for a deeper understanding of the workings of the protocols as well as for performance evaluation. We have also included socket programming taught at the undergraduate level to help students develop skills for network programming. Considering that students may have widely different backgrounds, we have included certain basics of networking to make the book self-contained. However, the introductory treatment on networking issues is given as a refresher of the basic concepts involved, and thorough learning of the relevant concepts from a networking book is a necessity before this book can be used.

It is well accepted that the knowledge acquired from a theoretical reading, especially in a subject such as computer networking, is incomplete when not accompanied by hands-on practice. The book is sprinkled with examples of simulations of both wired and wireless networking protocols. The assignments have been designed to be suitable for undergraduate and postgraduate levels of learning. For the advanced learner, suitable hints have been provided throughout the text to develop the skills for evaluation of new protocols.

Salient Features

(i) Emphasis on implementation and simulation of real-world network protocols.

(ii) Covers a wide ranging set of topics, starting from certain basic operating system commands to socket programming, wired network simulation, wireless network simulation, performance evaluation, and visualization.

(iii) Plenty of example programs have been provided (around ninety odd programs and scripts along with their explanations and outputs). Also many exercises (both theory and programming) requiring investigation and application of the learned concepts have been provided across all the chapters for practice.

(iv) We have tried our best to explain the concepts using simple language and analogies.

Content & Structure

Chapter 1 Discusses the evolution of data communication techniques and the fundamental issues associated with performance evaluation. It then provides an overview of simulation and other performance evaluation techniques.

Chapter 2 Introduces computer network protocols along with TCP/IP and OSI models. It also provides a brief overview of the networking devices used.

Chapter 3 Explains a socket and its use in network programming. This gives an idea of developing network applications using C and socket API.

Chapter 4 Introduces the NS2 network simulator, and exhibits the internal architecture of NS2 and its constituent software packages. It also provides pointers to installation of the package in different operating systems.

Chapter 5 Provides basic knowledge about simulation using NS2. It elaborates on the use of Tcl and OTcl scripts along with an introduction to AWK scripting and plotting with Gnuplot.

Chapter 6 Deals with the simulation of wired networks in detail. It shows how to simulate different protocols in different layers.

Chapter 7 Deals with the simulation of wireless networks in detail. It also discusses the idea of simulation, very large networks and measuring the various network parameters and plotting them in suitable graphs.

Acknowledgments

It is with great pleasure and pride that we thank all those who have given their unselfish help and support in many different ways in the preparation of the manuscript for this book. We thank our colleagues for their inspiration and valuable suggestions to improve the content and our staff members for their timely help. We are thankful to our students, as their feedback and comments helped to enrich the contents and to design new programs for the book. We are extremely thankful to our family members for their uncon-ditional and emotional support during the preparation of the manuscript. Finally, we thank the editorial team of Taylor & Francis Books India Pvt. Ltd. and in particular Dr. Gagandeep Singh for their constant support.

MATLAB® is a registered trademark of The MathWorks, Inc. For product in-formation, please contact:

The MathWorks, Inc.
3 Apple Hill Drive
Natick, MA, 01760-2098 USA
Tel: 508-647-7000
Fax: 508-647-7001
E-mail: info@mathworks.com
Web: www.mathworks.com

AUTHOR BIOGRAPHIES

Ajit Kumar Nayak is professor and head of the Department of Computer Science and Information Technology, Institute of Technical Education and Research, Siksha 'O' Anusandhan University, Bhubaneswar. He earned his M. Tech. and Ph.D. in computer science at Utkal University in 2001 and 2010, respectively. He has published more than 20 research articles in conference proceedings and journals. He is a member of IEEE and IET and a life member of the Orissa Information Technology Society (OITS). He has guided more than 15 M. Tech. students, and 5 research students are currently pursuing a Ph.D. under his guidance. His current research interests include mobile ad hoc networks, wireless sensor networks, and language computing.

Satyananda Champati Rai is currently working as an associate professor and head of the Department of Information Technology at the Silicon Institute of Technology, Bhubaneswar. He earned his M. Tech. (CS) and Ph.D. (computer science) from Utkal University in 2001 and 2012, respectively. He has published 20 research articles in national and international conference proceedings as well as in journals. He has written a monograph titled *QoS Provisioning in Mobile Ad Hoc Networks: Principles, Practices and Models*, published by LAMBERT Academic Publishing, Germany in 2013. He is a member of IEEE and a life member of the Orissa Information Technology Society (OITS) and ISTE. He has guided 22 M. Tech. students and served as a program committee member in several international conferences. His current research interests include mobile ad hoc networks, wireless sensor networks, and mobile cloud computing related to quality of service and performance analysis.

Rajib Mall earned B.E., M.E., and Ph.D. degrees from the Indian Institute of Science, Bangalore. He worked for nearly 3 years for Motorola India Ltd. before he joined the Department of Computer Science and Engineering, Indian Institute of Technology, Kharagpur in 1994 as a faculty member. He is currently a professor and the head of the CSE Department and the School of IT. His current research interests include program testing and software engineering issues in large systems and those in real-time embedded systems. He has published about 200 research papers and refereed international journals and conference proceedings and has carried out a number of sponsored projects in the areas of program analysis and program testing. He is a senior member of IEEE (USA).

CHAPTER 1

INTRODUCTION

The history of computers dates back more than 60 years. In the initial two decades, computers were operated largely as standalone machines, and primarily served as powerful number crunchers. About a decade later, the need for devising techniques to let applications running on different computers share information with each other was felt. This led to the birth of computer communication networks. An interconnected set of computers made it possible to develop powerful information sharing applications. This was in contrast to the plain number crunching applications that existed. This exemplified the potential advantages that computer communication networks can bring and caused the nascent computer communication technology to evolve at a rapid pace.

The mind boggling progress achieved in the area of computing technologies over the relatively short time span of the last six decades is now folklore. But even that pales when compared to the progress that has been achieved over the last half a century in the field of computer networking technologies. The dizzying speed with which computer networking technologies have advanced can be gauged from the fact that, almost every decade, the scope and contents of every networking textbook have been dramatically revised. In the following, we recount a few milestones in this evolution.

1.1 Rapid Evolution of Voice and Data Communication Techniques

In a groundbreaking work, Samuel Morse publicly demonstrated in 1838 that pulses of electric current can be used to move an electromagnet placed in a remote machine to produce dots and dashes on a piece of paper. The crude prototype demonstrated by Morse soon evolved into the telegraph system. The telegraph system revolutionized communications with far off places. However, with its inherent dot-and-dash Morse code mechanism, its use was largely restricted to communications of simple text messages. About 40 years later, in 1876, Alexander Graham Bell showed that voice signals can be transmitted as encoded time varying electric currents and then the current received at a remote destination can be used to reconstruct the voice signals. Bell's prototype was soon made into a simple telephone system that consisted of a pair of connected phone handsets. Such a simple telephone system could be used to connect a pair of fixed users. For example, a pair of phones could link two offices of a company. This was a network of size two (number of nodes is 2) and was obviously of very limited utility, as the utility of a telephone network is proportional to the square of the number of the nodes in the network. Shortly afterwards, on-demand connection establishment among different pairs of telephone handsets belonging to various subscribers was achieved through the use of manually operated exchanges. With the discovery of automatic telephone exchange technologies in 1879, circuit switching among different handset pairs became possible, and a subscriber could talk to other subscribers by simply dialing their numbers. This came to be known as the Public Switched Telephone Network (PSTN) system. A switch located in a telephone exchange could route calls to other exchanges. Efficient communication among subscribers residing in far away cities became possible with the invention of signal multiplexing techniques. Signal multiplexing made it possible to transmit multiple calls over the same trunk line. This led to more efficient usage of the physical medium.

Development of signal filtering techniques helped to reduce cross talk and other forms of noise in a phone call. The connection between two exchanges became known as a trunk line. The name trunk (analogous to a tree trunk) denotes that multiple communications are carried on the same line through use of either time division or frequency division multiplexing techniques. Initially, the telephone system carried only voice traffic and therefore was called a voice network. This completely analog technology is now called Plain Old Telephone Service (POTS) to differentiate it from the modern telephone systems, which, to a large extent, are carrying digital data with the help of many computing and electronic devices, rather than using analog devices such as filters, multiplexers, amplifiers, etc.

Wireless communication technology evolved independently almost over the same period of time as its wired counterpart. Wireless communication was born with the ground-breaking work of Guglielmo Marconi in 1899, when he made a bell ring at a remote place by pressing a button in the absence of any wired connections. Marconi's prototype was quickly commercialized, and it rapidly evolved into a host of technologies such as radio, wireless telegraphy, and wireless telephony.

1.2 Evolution of Computer Communication Networks

It was realized by many that the true power of number-crunching computers could be exploited by interconnecting them. They imagined that inter-computer communications could tremendously raise the practical utility of the number-crunching computers and convert them into processors of information or information systems. They visualized that an information system could gather data over geographically spread out locations and then process them to achieve useful results.

When the possibility of realizing technology that can enable computers to communicate was first investigated during the 1960s, researchers naturally tried to make use of the telephone networks that already existed at that time (the Plain Old Telephone service (POTS) networks) for carrying data. However, this effort encountered two major obstacles:

- First, computers communicate digital data among each other and not analog voice signals. The problem caused by this fact concerned the following: "How can digital data be transmitted over POTS networks?" This was a relatively easy problem to solve, as modems could be developed to encode digital data over an analog carrier signal that could efficiently be transmitted on a telephone network.

- The second problem was more severe. POTS networks used circuit switching to establish connections among pairs of users. Circuit switching involved allocating a fixed amount of bandwidth to each connection. On the other hand, transport of bursty traffic is the hallmark of all computer data communications. Reserving bandwidth for bursty traffic clearly leads to a huge waste of bandwidth. Packet switching is much more advantageous for communications involving bursty traffic. However, so much investment had already been made in setting up POTS networks that the potential advantages that packet switching could offer had to be ignored. Packet switching could not be implemented for a long time, as it would have required immediate overhauling of the existing circuit-switched exchanges into which billions of dollars of investments had already been made.

The inefficiency of POTS networks in carrying data traffic caused the inherently efficient data networks (also known as computer communication networks) to independently and speedily evolve. It may be worthwhile to note that initially computer communication networks carried alphanumeric data only and hence were called data networks, whereas POTS networks carried voice information only and hence were called voice-oriented networks. Data networks no longer carry alphanumeric ASCII data, but carry various types of data such as images, encoded information, etc. Therefore, these can be more appropriately called bit streams. However, the name data networks has stuck. In later generation networks, voice and data networks are slowly becoming indistinguishable for reasons we discuss later.

It is important to know the key ways in which data networks differ from POTS networks. A principal way in which they differ is that a POTS network carries analog signals and uses circuit switching effected by exchanges, whereas data networks carry digital data in packets and packet switching is provided by networking devices such as hubs, switches, routers, and gateways.

Table 1.1: Landmark events in the evolution of computer communication networks

Late 1960s	First data communication over voice telephone networks
Late 1970s	Development of TCP/IP and design of the Internet
Early 1980s	Commercial usage of Ethernet
Late 1980s	Commercial usage of Internet
Early 1990s	First usage of World Wide Web
Late 1990s	Usage of wireless data networks

Data networks and protocols evolved in a much more remarkable way compared to POTS networks. In Table 1.1, we have summarized the landmark events in the evolution of data networks. As already pointed out, the first computer data communications over telephone networks occurred in the early 1960s. This was achieved by modulating digital data on a carrier wave form. Another major milestone was reached when TCP/IP was developed as part of the ARPANET project of the US Department of Defense (DoD), which wanted to develop technologies to connect dissimilar networks that were spread over a large geographic area.

Initially, TCP was called Transmission Control Program and was written as a single program. Later, in March 1978, TCP was split into two different protocols: Transmission Control Protocol (TCP) and Internet Protocol (IP). This splitting of TCP into TCP and IP became necessary because, as far as streaming data is concerned, fast transport of data could be achieved by inhibiting error checking and this could be provided through the design of the User Datagram Protocol (UDP). The splitting of TCP made it possible to use UDP over IP.

TCP and IP are considered to be the core protocols of the TCP/IP suite. Later, several other protocols were subsequently added to the suite as and when the need for them arose. For example, FTP (File Transfer Protocol) was developed when the need for transferring large files over a network arose. In the early 1990s, the world wide web (WWW) emerged and made TCP/IP the de facto protocol of the Internet. The Internet is possibly the most important example of a data network.

As shown in Table 1.1, development of wireless networks and later the Ethernet are landmark events in the history of the development of computer communication networks. Norman Abramson, who was a professor at the University of Hawaii, is credited with the development of ALOHAnet in the 1970s. His fundamental work on ALOHAnet led to the development of wireless networks, and subsequently to the development of Ethernet in the late 1980s. A few years later, cellular wireless networks which initially supported transmission of voice signals came into being. Later these evolved to support data networks with the introduction of 3G cellular telephony in the early 2000s.

1.3 Convergence of Data and Telecommunication Networks

It was clear that circuit-switched telecom networks are a poor fit to bursty computer communication traffic and therefore perform poorly. However, so much was invested in setting up telephone networks across the world that it was initially not considered worthwhile to discard or drastically redesign them. Rather, development of technology for efficient data transmission over the existing telephone networks was attempted. A prominent attempt to support data communication over the existing telephone networks was made through the development of ISDN (Integrated Services Data Network). ISDN stands for "integrated services" since digital and analog transmissions occur over the same wire. In fact, ISDN is based on having multiple channels over the same regular telephone line. In ISDN, computers can directly transmit digital data over the telephone network without having to first convert digital data into an analog audio signal using a modem. At present, ISDN is being used extensively for providing many types of services over telephone lines, including data services for linking bank ATMs, fax, etc.

Modern data networks now use a variety of physical media: twisted pair, fiber optic, wireless, etc. The different network segments are interlinked by hubs, switches, etc. It has now been well accepted that, even for pure voice transmissions, circuit switching is inefficient. This is because, over a moderately long interval, voice signals tend to have long periods of silence. Holding up the line for the entire duration of a conversation, as done in circuit switching, therefore is obviously inefficient. Data networks primarily use packet switching and have been designed for supporting efficient computer data communication. Due to the efficiency and rapid expansion of data networks, analog voice signals are nowadays increasingly being transmitted as digitally encoded packets over data networks, to take advantage of the inherent efficiency of packet-switched networks. This is slowly making convergence of data and voice networks a reality. The rising popularity of voice over IP (VOIP)-based applications bears testimony to this.

1.4   Integration of TCP/IP into Unix

Unix was developed by AT&T Bell Laboratories in 1972. However, the business focus of AT&T was telecom and it was not interested at that time in commercially exploiting Unix. Consequently, Unix was made into an open source software application and was given away free of cost to any organization that requested it. The University of California, Berkeley (UCB) was one of the early recipients of the Unix source code. DARPA (Defense Advanced Research Projects Agency) awarded a large project grant to UCB to program TCP/IP into Unix for supporting easy communication among different Unix-based computers. In this regard, UCB proposed the socket interface, which a programmer could use to access the TCP/IP implementation in Unix. UCB's version of Unix came to be known as BSD (Berkeley Software Distribution). Now, sockets have become a standard feature of all Unix distributions and have been made a mandatory feature for operating systems by the POSIX standard. In the following subsection, we briefly explain the POSIX standard at the risk of digressing from the core theme of this text.

1.4.1   POSIX

POSIX was started as an attempt to standardize various flavors of Unix operating systems. POSIX stands for Portable Operating System Interface. It was proposed in the context of Unix operating systems. Therefore, the letter "X" was used to make it sound Unix-like. POSIX started as an open source software initiative, but over the last two decades it has become a widely accepted and de facto standard for operating systems, including real-time operating systems. The importance of POSIX can be gauged from the fact that nowadays it is rare to come across a commercial proprietary or open source operating system that is not POSIX compliant. In the following, we briefly recount the genesis of POSIX.

As already mentioned, Unix was originally developed by AT&T Bell Labs in the early 1970s. Since AT&T was primarily a telecommunications company, it felt that Unix was not commercially important for it. Therefore, it distributed Unix source code free of cost to any one interested. UCB (University of California at Berkeley) was one of the earliest recipients of Unix source code.

AT&T later got interested in computers. It realized the potential of Unix and started developing Unix further and came up with Unix V. Meanwhile, UCB had incorporated TCP/IP into the Unix source code through a large DARPA (Defense Advanced Research Project Agency of the USA) grant. Subsequently, UCB came up with its own version of Unix and named it BSD (Berkeley Software Distribution). At that time, the commercial importance of Unix had started to grow very rapidly. As a result, many vendors extended the free source code of Unix in different ways and came up with different versions of Unix. A few prominent examples are AIX from IBM, HP-UX from

HP, Solaris from Sun Microsystems, Ultrix from Digital Equipment Corporation (DEC), and SCO-Unix from SCO.

Since so many variants of Unix from different vendors came into use, portability of applications across different Unix platforms became a major issue. It resulted in a situation where a program written on one Unix platform would not run on another Unix platform. In this context, the main objective of POSIX was to facilitate portability of applications across different operating systems. The POSIX standard defines only the syntax of the interfaces to operating system services and the semantics of these services, but does not specify exactly how the services are to be implemented. In simple words, POSIX standardizes the syntax (names of the system calls, their parameters, etc.) and the semantics (what is achieved by making a system call), but does not restrict exactly how a system call is implemented.

1.5   Queueing Theory

As we have discussed, there are several ways of evaluating the performance of systems. Application of queueing theory is an important one. Whenever the inputs to a system as well as the behavior of the system are deterministic, analytical evaluation of the performance parameters of the system is usually straightforward. Analytical solutions are accurate and easy to automate. However, many systems that we encounter in real life have inputs that are statistical in nature and the system as a whole or the different components of the systems may show statistical behavior. Examples of such systems are computer networks, customers turning out to withdraw cash at a bank ATM, voters turning out to cast a vote at an election booth, patients turning out for treatment at a hospital, etc. Queueing theory is a popular way to determine the performance of systems with statistical behavior, such as packet transmission in a computer network. In fact, queueing theory to a large extent owes its rapid development to its successful use in telecommunication systems. Since then, it has been successfully used for performance evaluation of computer networks, computer systems, machine plants, etc. Queueing theory first constructs a simple model of a system and then evaluates the performance of the system based on this model. Queueing theory typically models a system as one or more servers providing service of some sort to the arriving inputs. A simple model of a queue is shown in Figure 1.1.

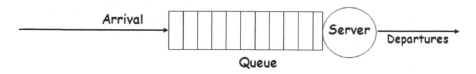

Figure 1.1: A simple model of a queue

In our everyday life, we encounter queues formed at many places. As an example, we consider a salon. Customers come to a haircutting salon for a haircut. A customer who arrives when the number of customers in the system is low gets service immediately from a server. So a customer's time in the system (called residence time) is essentially equal to the service time. However, when the number of customers who are already in the system is large, an arriving customer finds all servers busy and has to wait in a queue. Since "customers arriving for a certain service" is a very intuitive model for a queue system, we shall at times use the term customer for input to simplify the explanations. From our above discussion, it is easy to see that a queueing system models a statistical system in terms of three components:

- Arrival process

- Service mechanism

- Queue discipline

In the following, we discuss these components of a queueing system briefly.

Arrival Process
The arrival process describes how the inputs arrive in the system. The time that elapses between two successive arrivals of inputs to the system is called the interarrival time. If Ai is the inter-arrival time between the arrivals of the $(i-1)$th and ith inputs to the system, then the mean (or expected) interarrival time of the inputs is denoted by E(A). From this, the arrival frequency can be given by $1/E(A)$, which is also denoted by λ. In many practical systems, the interarrival times are independent and follow a certain statistical distribution. A Poisson distribution is very common and denotes exponential interarrival times.

Service Mechanism
The service mechanism of a queueing system is usually specified by the number of servers present in the system. Each server may have its own queue, or there can be a common queue in which the customers can wait for any of the servers. Besides the number of servers, a service mechanism is characterized by service times. Typically, service times of different customers are independent and follow some statistical distribution. Let Si be the service time for the ith customer. We denote the mean service time of a customer by E(S) and $\mathfrak{t} = 1/(E(S))$ is called the service rate of a server. It is usually assumed that the service times are independent of the interarrival times. However, it may be possible that, in some situations, the service times are dependent of the queue length. For example, when the number of customers in a salon is too large, the haircutter may become tired and take an increasingly long time to serve customers.

Queue Discipline

Queue discipline is also called service discipline. It defines the order in which customers waiting in a queue are served. Once customers have queued up, there are many possibilities for the order in which they can be served. Some of these possibilities are the following:

- FCFS: First come, first served.

- Random order.

- LCFS: Last come, first served.

- Priority order, etc.

Measures of Performance for Queueing Systems

As we have already mentioned, a major objective of queueing theory is to evaluate the performance of a statistical system. The performance of the system under consideration needs to evaluated in terms of certain performance measures. There are many possible measures of performance that are in common use. Only some of these will be discussed here. We first discuss steady state average delay and the steady state average waiting time in the system. Let D_i be the delay in queue for the ith customer before he is served and W_i be the total time that the ith customer spends in the system. Obviously, the total time that a customer spends in the system is the sum of his delay before being served and the service time. In other words, $W_i = D_i + Si$.

$$d = \lim_{x \to \infty} \frac{\sum_{i=1}^{i=n} D_i}{n}$$

and

$$w = \lim_{x \to \infty} \frac{\sum_{i=1}^{i=n} W_i}{n}$$

d and w are called the steady state average delay and the steady state average waiting time in the system. Let $Q(t)$ be the number of customers in the queue at time t. $L(t)$ be the number of customers in the system at time t. Obviously, $L(t) = Q(t) +$ no. of customers being served at t.

Q and L are called the steady state time average number in the queue and the steady state time average number in the system. A queueing system is usually described using Kendall's notation. A queue system can be characterized by Kendall's notation, whose general form is $[A/B/X]:\{D/Z\}$ where

A = Probability distribution of the arrivals
B = Probability distribution of the departures
X = Number of servers
D = The maximum capacity of the queue(s)
Z= The queue discipline

From Kendall's notation, the following can be inferred. An [M/M/1]:∞/ FCFS system is one where the arrivals to and departures from the system follow a Poisson distribution, there is a single server, the queue length is infinite, and the queue discipline is FCFS. This is the simplest queue system that can be studied mathematically. This queue system is often referred to as the M/M/1 queue.

The following aspects are used to characterize a queueing system.

(i) The input (arrival pattern): The input describes the way in which customers arrive and join the system. It is typically specified in terms of the mean arrival rate λ and the statistical distribution for arrival. We shall mostly deal with a Poisson arrival pattern.

(ii) The service mechanism: The distribution of service time is an important characteristic of a queueing system . The mean service rate is denoted by μ.

(iii) The queue discipline: This is a rule according to which customers are selected for service when a queue has been formed. We have already mentioned that the most common disciplines are FCFS, random, LCFS, and priority-based.

(iv) Customers' behavior: Customers generally behave in the following ways.

- Balking—A customer may leave the queue because the queue is too long or the queue is full.

- Reneging— A waiting customer may leave the queue due to impatience.

- Jockeying— A customer may jockey from one queue to another.

Probability Distribution of Arrivals and Departures

The probability distributions that are popularly used to describe arrivals and departures are the following:

M = Poisson (Markov) process
E = Erlang distribution
G = General distribution

Poisson Process

Poisson process is possibly the most popular arrival and departure model used in queueing theory. A Poisson process is the stochastic process in which events occur continuously and independently of one another. Examples that are well modeled as Poisson processes include the number of packets arriving at a switch, the page view requests to a web site, the number of people arriving at an ATM booth to withdraw money, etc. The Poisson process can be described in three different, but equivalent ways.

1. A Poisson process is a pure birth process. In an infinitesimal time interval dt there may occur only one arrival. This happens with probability $\lambda\, dt$ independent of arrivals outside the interval.

2. The number of arrivals $N(t)$ in a finite interval of length t that obeys the Poisson (λt) distribution. That is, the probability of n arrivals over a duration of t units of time is given by
$$P_r[N(t) = n] = \frac{(\lambda t)^n}{n!} e^{-\lambda t}$$

3. The interarrival times are independent and follow the exponential distribution. That is, P_r[Interarrival time$>t$]]

Therefore, the probability of zero arrival in the interval $[0,t]$ is P_r(zero arrival in $[0,t]) = e^{-\lambda t} = p(0)$.

M/M/1 Queue
This makes the following assumptions.

(i) Arrivals follow a Poisson distribution

(ii) Service time follows an exponential distribution

(iii) Single server

(iv) Infinite capacity

(v) First come first served service discipline

Little's Theorem
Little's theorem is a fundamental result that holds true for all arrival and service processes. It states that the mean number of customers in a queueing system is the product of their mean waiting time and their mean arrival rate.

Markovian Systems
The Markov process named after the Russian mathematician Andrey Markov. A Markov process is "memoryless" in the sense that one can make predictions for the future of the process based solely on its present state. For a continuous Markov chain, the process may remain in any given state for a time that is exponentially distributed. A birth-death process is a very special class of Markov processes where the state transitions are only to the neighboring states. In an M/M/1 queue, all distributions of the system, the service times, and the interarrival times are independent of one another or the state of the system (memoryless) and are exponentially distributed. Thus queueing systems with Poisson input can be considered Markovian processes. All Markovian processes can be mapped to a continuous time Markov chain (CTMC), which can be easily solved.

In Figure 1.2, a continuous time Markov model of an M/M/1 queue is shown. Observe that the state of the system is defined (and also labeled) by the number of customers in the queue. The arrival of a customer occurs at an average rate of λ and departure takes place at an average rate of ρ. The

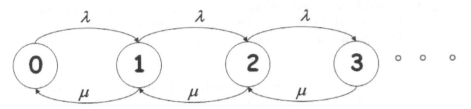

Figure 1.2: Continuous Time Markov Model of an M/M/1 Queue

system enters a state to the right on the arrival of a customer and enters a state to the left on the departure of a customer.

Performance Measures for an M/M/1 Queue

Based on the continuous time Markov model of an M/M/1 queue, several performance measures for the queue can be determined. Let $\rho = \frac{\lambda}{\mu}$. This term is called the traffic intensity. An M/M/1 queue is stable iff $\lambda < \mu$ or < 1. For a stable M/M/1 queue, the following performance measures can be derived.

a) Mean number of customers in the system $L = \rho/(1 - \rho)$

b) Mean time to go through the system $= 1/\mu \, (1 - \rho)$

c) Mean waiting time for a customer in the queue (W_q) = $\rho/\mu \, (1 - \rho)$

d) Mean number of customers in the queue $= \rho^2/(1 - \rho)$

e) Probability that the queue is not empty (server is busy) $= \rho$

Example 1. The time that a haircutter takes to cut the hair of a customer is an exponential distribution with a mean of 30 minutes. If he provides his services to customers in the order in which they came in, and if the arrival of customers follows a Poisson distribution with an average rate of 10 per 8 hour day, what is the haircutter's expected idle time each day? Is a customer expected to find how many are waiting ahead him?
Solution
$\lambda = 10/8 = 5/4$ customers/hr
$\mu = (1/30) \, 60 = 2$ customers/hr
$\rho = 5/8$
Number of hours the haircutter remains busy in an 8-hr day = $8*\rho = 8(5/8)$= 5 hr
Haircutter's expected idle time in an 8-hr day = 8–5 = 3 hr
Expected number of customers waiting for the haircutter = L = $\rho/(1 - \rho)$ = (5/8)/[1–(5/8)] = 5/3 = 2 sets (approx.)

Example 2. Consider a network router. Suppose that the Poisson arrival rate of packets for execution is 4 packets/sec and the average (exponentially distributed) service time is 0.20 sec.

a) What is the probability that an arriving packet will have to wait before receiving service?

b) What is the average amount of time a packet spends waiting to be serviced?

c) What is the average delay that a packet experiences?

Solution:
$\lambda = 4$ packets/sec
$\mu = 5$ packets/sec
$\rho = 0.8$

a) The probability that an arriving packet would have to wait before receiving service = 0.2

b) Amount of time a packet spends waiting to be serviced = ρ/μ $(1-\rho)$ = 0.8/(5*0.2) = 0.8 sec. Average delay that a packet experiences = $1/\mu$ $(1-\rho)$ = 1/5*0.2 = 1 sec

1.6   Overview of Simulation

A simulation experiment, in essence, involves performing an imitation of the operations of a real system such as a network protocol over a certain period of time with the objective of evaluating the behavior of the actual system. A simulation experiment can help to record the imitated behavior of the system over a specified time interval. The recorded information can then be analyzed to draw meaningful conclusions to gain insights into the behavior of the actual system. In general terms, simulation involves designing a model of a system to be studied and then experimenting with this model to study the behavior of the system over some predefined time period. From this definition, we can conclude that modeling is inherent to any simulation.

Any model can be considered a simplistic representation (or an abstraction) of the actual system. Consequently, a model can help in quick evaluation of the behavior expected of an actual system in a realistic environment. On the other hand, it can easily be seen that simulation results are usually gross approximations of the behavior of the actual system. This is due to the fact that a model is only a simplified representation of the actual system. Any simulation using a model therefore produces approximate behavior of the actual system and leaves out many aspects of the system's detailed behavior. In the following, we discuss the pros and cons of carrying out simulation studies over experimenting on a real system in a real deployment environment.

1.6.1   Advantages of simulation

The following are the important advantages of carrying out simulation studies of a system as compared to experimenting directly on the actual system.

- **Helps to evaluate alternate designs when the real system is not available:** Possibly the greatest strength of simulation is the ability to evaluate the performance of the actual system, when the actual system has
  not yet been constructed. For example, during the design stage, the system is far from being constructed. But the designer may wish to evaluate the impact of various design options on the performance of the
  actual system. For example, a network protocol design team may design alternate network protocols for a specific application environment
  and wish to determine the superior one. Simulation experiments therefore can help to answer the "what if" questions. The use of appropriate
  simulation techniques has been accepted as a promising way to arrive
  at better design solutions.

- **Less expensive and faster:** Field tests using the actual system are usually expensive. For example, consider studying the performance of a
  specific protocol in a sensor network having different types of sensors
  (pressure, temperature, motion, etc.) that generate traffic of different
  characteristics. Now suppose the efficacy of a few alternate designs of
  a network protocol in such a network needs to be studied. The different steps required to be carried out for this study would include setting
  up the network, implementing the proposed protocol variants, generating the required traffic by the sensor nodes, monitoring and recording the traffic at various points in the network, etc. Therefore setting
  up these steps for a real system for experimentation can be pretty expensive as well as very time consuming. Not only is experimenting
  using real systems prohibitively expensive, it can even be dangerous.
  Consider a novice trying to learn to fly a real air plane as compared to
  using a flight simulator.

- **Easy to reproduce any specific observation:** Experimental evaluation
  of the performance of a complex system can be hard to carry out and
  therefore any observed anomalous behavior is often difficult to reproduce. For example, the results of a networking experiment may depend
  on the specific device settings made. Such settings may be time consuming and laborious to carry out, and any mistake in the exact device
  setting made after the settings have been changed for a later experiment can make it hard to reproduce the behavior observed in an earlier
  experiment.

- **Helps to evaluate alternative configurations without disrupting a real
  system:** Simulation can be used to study the impact of a change to a
  working system without disrupting ongoing operations. For example,

the specific configurations that may be most suitable in a real network may first be evaluated through simulation rather than inconveniencing users by trying out the different possible settings on the live network and then taking measurements on this.

- **Helps to improve understanding of a system:** Simulation experiments can help to identify bottlenecks and to gain insights into the performance of a complex system such as a commercial network. For example, they may help to determine which parameters play a crucial role to achieve high system performance and what should be the optimal values for these variables for the best achievable network performance.

- **Study of stochastic systems:** Closed-form mathematical solutions to stochastic processes such as arrival of packets at different routers with different traffic probability distributions for nontrivial situations can become overly complex. Experiments involving stochastic systems can take an arduously long time and controlled experiments may be difficult to carry out on real systems that are in use. On the other hand, solutions to mathematical models of such stochastic systems would require many simplifying assumptions to be made, leading to inaccurate results. Simulations are an important way to arrive at meaningful and fairly accurate results in the case of stochastic systems. Meaningful experiments on the performance of computer communication networks are usually stochastic in nature.

- **Allows us to control time:** Using simulation, the long-term behavior of a network with various network traffic and topology patterns can be studied in a relatively short time interval with proper setting of the values of the time parameters. For example, if we wish to study the behavior of a network protocol spanning a few days in a matter of a few minutes, it becomes possible by setting the time granularity appropriately in a simulation experiment.

1.6.2   Disadvantages of simulation

While simulation has many advantages and provides a quick insight into the performance of an existing or a proposed solution, it does have certain disadvantages which anyone relying on simulation results must bear in mind.

- **Difficulty in building an appropriate model:** A basic step in any simulation study is model building. However, model building is an art as well as a science. Model building is an art because one has to use subjective judgment to decide which aspects or parameters are crucial to a given behavior of the system and ignore the rest. Model building is a science because modeling techniques have precise syntax and semantics, and systematic inferences can be drawn using them.

In other words, a large number of correct models of different granularity (i.e., different level of abstraction) designed for the same system yield results of widely varying accuracy. However, the objectives of a simulation experiment can be best met only by a few models of the right granularity from the very large number of models that may be possible. Building the right model requires significant expertise in both qualitative and quantitative aspects of model building. In other words, the quality of simulation results depends on the quality of the developed model, which in turn largely depends on the skill of the modeler.

- **Time and effort intensive:** Simulation is an iterative process and therefore simulation experiments can be both time consuming and expensive to carry out. Developing simulation models and their implementation can take several months, and large-scale simulations may require extremely large amounts of computing power and time. Further, data produced by a simulation experiment can be extensive, and analysis of the simulation results might require significant effort.

- **Simulation results can be misleading:** Simulation experiments may lead to nonintuitive and even wrong results unless one has good modeling skills, develops knowledge of the nitty gritty of performance evaluation techniques, and develops familiarity with the prevalent simulation tools.

- **Hard to verify the simulation results:** Verification of the correctness of a simulation model as well as the simulation results is often hard. Therefore, unless sufficient care is taken, it is possible to arrive at wrong conclusions through simulation experiments due to construction of a wrong model or a bug in the simulation software.

1.6.3 Types of simulation

There are many ways to categorize existing simulation techniques. A common classification of a simulation technique is based on whether it is (i) static or dynamic, (ii) stochastic or deterministic, (iii) a discrete event or continuous. In the following, we briefly discuss these three classes of simulations.

Static versus dynamic simulation: Static simulation, as the name implies, applies to systems for which the computed results do not vary with time for a given experiment. A static simulation is applicable to systems whose behavior is independent of time. That is, time is not a parameter in the behavior of these systems. Thus, conceptually, we can consider such a system as accepting a few inputs and producing outputs that are invariant with time. The evaluation of such systems often involves generating random input values to generate a statistical outcome. An important example of a static simulation technique is the Monte Carlo simulation. We shall discuss this technique in some detail in Section 1.7.1.

Dynamic simulation involves simulating a system over some specified time interval. Obviously, dynamic simulation is applicable to systems whose behavior may vary with time, possibly due to changes in the state of the system. Dynamic simulation can help consider the changes to the state of a system and the corresponding change in system behavior as they occur over time. Computer networks require dynamic simulation because the behavior of the system changes as packets are generated at different rates by the traffic sources, are queued at the routers, and are delivered to the appropriate destinations. Since keeping time is a vital aspect of dynamic simulation, a software clock is typically used for this purpose. There are two main ways in which the clock time is incremented. The clock time is either advanced in fixed increments in each iteration of the loop or is advanced based on the events that occur. Based on the way the clock time is incremented, the simulation of a dynamic system is called either clock driven or event driven. In both types of simulation, the termination of the simulation is usually based on the clock value reaching a certain reading.

Stochastic versus deterministic simulation: In a deterministic simulation, the model does not contain any probabilistic components. A deterministic system is usually modeled as a set of differential equations or in terms of linear regression. On the other hand, dynamic simulations involve some randomness in behavior and are referred to as stochastic or probabilistic simulations. For example, one or more input variables may have to be chosen randomly, as is usually the case in a network simulation experiment. Arguably, stochastic simulations are more useful, as these help to obtain nontrivial insights into the behavior of a dynamic system. This is possibly because these take into account the uncertainty due to the varying behavior of a system to different inputs. A stochastic simulation experiment produces output that is itself random, and therefore each experiment gives only one sample data point as to how the system might behave (behavior space). Therefore stochastic simulation experiments must be performed a number of times to get a realistic evaluation of the system.

As already pointed out, stochastic simulations are run using random inputs, and these produce random outputs. Performance estimates for stochastic simulations are obtained by calculating the average value of the performance metric and the deviation computed across a large number of runs. If conclusions need to be drawn based on a limited number of runs, then confidence intervals must be used. We discuss some basic concepts in constructing confidence intervals in Section 3.2. In contrast to stochastic simulations, deterministic simulations need to be run only once to get precise results because the results will always be the same.

Discrete-event versus continuous-time simulations: A discrete-event simulation is one for which the state variables change only at certain points in time that are defined by an occurrence of an event. A continuous-time simulation, on the other hand, is one for which the state variables change continuously with respect to time. Discrete-event simulations are suitable for problems in which the system state variables change at discrete times and in discrete steps. A large number of practical systems fall under this category. Computer networks and most manufacturing and service systems are examples of discrete-event systems. On the other hand, in a continuous-time system, the change to the system state variable may occur continuously over time. We can say that, in discrete-event simulation, the occurrence of events drives the simulation, whereas in continuous simulations the passing of time drives the simulation. In the following, we illustrate the difference between continuous and discrete simulation using an example.

Consider an oil tanker at a port loading oil. The state variable in this case is the volume of oil in the tanker. If the tanker is loaded by transporting several smaller filled containers onto it, then it is a discrete-event system. In this case, the state variable (volume of oil in the tanker) assumes only a discrete set of values and the system state variable changes at discrete points of time. On the other hand, if the oil tanker is loaded by pumping oil into it through a pipeline, then it is a continuous-time system and the system variable changes continuously. The difference between a discrete-event and a continuous-time system is brought out in the plot in Figure 1.3. For the oil tanker loading example, the plot in Figure 1.3 shows how the system state variable (volume of oil in the tanker) changes over time. The bar chart represents the value of the state variable for the discrete-event system. Notice that, at certain discrete points of time, the smaller containers are loaded, leading to discrete increments to the state variable. On the other hand, in a continuous-time model the state variable varies continuously with time. Continuous simulation requires the use of either difference equations or differential equations to define the rate of change of the system variable. A detailed discussion of this is out of the scope of this text.

1.7   A Few Basic Concepts in Simulation

In this section, we first discuss the Monte Carlo simulation, a static and statistical simulation technique. This technique was one of the early simulation breakthroughs. Manual execution of Monte Carlo simulation is effort intensive and error prone. The true power of Monte Carlo simulation could be exploited with the easy availability of computers in the 1980s, as Monte Carlo simulations could be programmed, leading to automated execution. First, we discuss certain important issues in Monte Carlo simulation. We discuss the evolution of computerized simulation techniques and review a few basic concepts in computerized simulation.

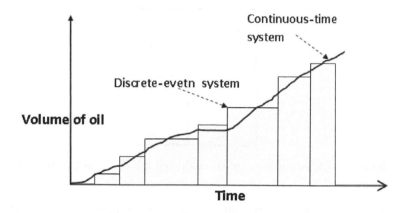

Figure 1.3: Continuous-time-event versus discrete systems

1.7.1   Monte Carlo simulation

An important breakthrough in simulation occurred during World War II in the form of the development of "Monte Carlo" simulation techniques. The name possibly arose from the analogy that it is based on using randomly generated numbers, as in the games played at the famed casinos at Monte Carlo. However, this similarity in the use of random numbers is very shallow, and a more appropriate name could have been stochastic simulation. This simulation technique is still widely used today for problems which are not analytically solvable. It is a static simulation technique and cannot handle time-varying (dynamic) systems. It keeps on generating random numbers and computing with these random input values until some equilibrium state is reached. Therefore the process is inherently iterative, and a significant number of iterations must be performed before results with sufficient accuracy can be obtained. Obviously, for system models requiring complex computations for producing the output from the input, manual Monte Carlo simulation becomes tedious. Therefore this technique has become much more popular with the advent of computers.

Monte Carlo simulations have been acknowledged to be an effective approach to solving problems that are complex, nonlinear, or involve more than several uncertain parameters. As already mentioned, Monte Carlo simulations involve developing a mathematical model of the system and then repeatedly generating random numbers to compute the results. Since this technique is based on repeated computation of the model expressions by using random numbers, it is rather difficult to use this technique to manually simulate nontrivial systems. It becomes easier to use this technique when run on a computer as a computer program. Monte Carlo simulation has been used to investigate physical phenomena, evaluate various probabilistic systems,

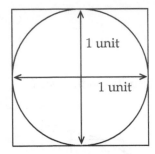

Figure 1.4: Determination of the value of π by Monte Carlo simulation

as well as numerically estimate the value of complex mathematical expressions for which closed-form solutions are not available. In the following, we explain the use of the Monte Carlo technique through an illustrative example.

Example 1.1: Determine the value of π by using Monte Carlo simulation.

Solution: Consider the circle shown in Figure 1.4 whose radius is 1 unit. A square has been superscribed on the circle. The diameter of the circle and hence each side of the square would be 2 units. The area of the circle would be $\pi \times 1^2 = \pi$. Therefore the area of the square is 4 square units. The ratio of the area of the square to that of the circle is $R_{sc} = \pi/4$. Now imagine throwing a dart at the square. This can be considered to be similar to generating a random point whose x and y coordinates are random numbers, each in the range $[0, 2]$. Now, we would count a point to be inside the circle if $\sqrt{(x^2 + y^2)} < 1$. We can compute the ratio R_{sc} of points inside the circle and those outside it. That is, $R_{sc} = in/(in + out)$, where in is the number of points lying inside the circle and out is the number lying outside it. After computing R_{sc} by generating a large number of points, the value of π can be given as $4 \times R$.

1.7.2 Confidence intervals

Suppose there are 1000 students (called the population) taking an examination and you happen to know the grades of 20 of them (called the sample). Obviously, you cannot claim that you know the average grade for all 1000 students. However, you can compute the average of the 20 grades that you know. But the average computed from the 20 grades that you know may be very different from the average computed by considering the grades of all 1000 students. Still, it would be possible to conclude certain meaningful information about the average grades scored by all the students by using confidence intervals.

The 20 grades that you know is called a sample. More formally, a sample is any finite subset of elements chosen from the population. A point estimate is any observation made based on the sample. A point estimate, by itself, conveys very little information because it does not reveal the uncertainty associated with the estimate. For example, you cannot even say by how much the mean computed from the sample differs from the actual mean. For example, suppose you computed the average grades of 20 students to be 45. Suppose you are asked whether you are confident that the actual population mean lies within 5 grades of 45. Of course, you cannot be sure. *The confidence intervals are useful in these situations and provide more information than point estimates.*

Confidence intervals are central to making meaningful inferences from the results of stochastic simulation experiments. During a simulation experiment, we attempt to generate only limited samples of a large number of representative scenarios for a system. Complete enumeration of all possible scenarios would be prohibitively expensive. For these experiments, the use of confidence intervals provide a mechanism to infer useful information regarding the overall behavior of the system.

Suppose, for a given population, we compute the population mean (μ) and then take several samples of this population. We can then define computation intervals on the sample means as follows. Suppose we define the intervals (called confidence intervals) around the sample mean such that these contain the population mean (μ) a specified proportion of the time, say 95% or 99% of the time. These intervals are referred to as 95% and 99% confidence intervals, respectively. A 95% confidence interval implies that there is a 95% chance that the confidence interval computed for a sample contains the population mean (μ). As an example, consider the following 95% confidence interval: $72.8 < \mu < 97.1$ This confidence interval means that, if several samples are taken and confidence intervals computed for each of them, then the intervals computed for 95% of the samples contain the population mean. Naturally, the intervals computed for 5% of the samples do not contain the population mean.

We can now define the 95% confidence interval as the one that has a 95% probability of containing the actual population mean (μ). A confidence interval such as $72.8 < \mu < 97.1$ is usually denoted as $(72.8, 97.1)$. We now briefly discuss how the population mean can be computed from the sample mean.

For each experiment, a different mean value is obtained. By the central limit theorem (CLT), the mean computed for different samples ($\bar{x}$) behaves as a normally distributed random variable with its mean equal to the actual population mean (μ). That is, the mean of all the sample means is the same as the mean of the population from which the samples are taken. Note that the distribution of the sample means is normal regardless of the exact population distribution.

The standard deviation of the sample mean is not the same as the standard deviation of the population the sample is taken from, but it is related to it by $\sigma_{\bar{x}} = \frac{\sigma}{\sqrt{n}}$, where $\sigma_{\bar{x}}$ is the standard error of the mean. Notice that it gets smaller as the sample size (n) grows. This is intuitive. The bigger the sample size, the more likely the sample mean is closer to the true mean. If the sample size were equal to the whole population, it is in fact the sample mean.

In summary, the CLT tells us that, for a large enough sample, the sample mean is normally distributed with a mean of μ and a standard deviation (standard error) of $\frac{\sigma}{\sqrt{n}}$. Let us now formalize our discussions regarding point estimates and confidence intervals.

Point estimate: The value computed from a sample statistic is called a point estimate. Usually, whenever we use point estimation, we calculate the margin of error associated with that point estimation.

Confidence interval estimation: This involves constructing an interval around the point estimate, such that this interval is likely to contain the corresponding population parameter. The confidence level associated with a confidence interval states how confident we are that this interval contains the true population parameter. The confidence level is denoted by $(1 - \alpha)100$, where α is the chance that it does not contain the true mean. For example, if $\alpha = 0.05$, then the confidence level is 0.95, or 95%, and we say that we are 95% confident that the interval contains the mean. From established statistical results, we reproduce the following:

The 90% confidence interval is given by $\bar{x} \pm 1.645 \frac{\sigma}{\sqrt{n}}$

The 95% confidence interval is given by $\bar{x} \pm 1.96 \frac{\sigma}{\sqrt{n}}$

The 99% confidence interval is given by $\bar{x} \pm 2.58 \frac{\sigma}{\sqrt{n}}$

From the above, we can observe that 95% of the sample means lie within $\bar{x} \pm 1.96\,\sigma_{\bar{x}}$ of the population mean. We can also state that any arbitrary sample mean has a 95% chance of being within $1.96\,\sigma_{\bar{x}}$ from the population mean. Also, observe from the above set of expressions:

- Larger samples lead to lower standard errors and smaller confidence intervals

- Larger standard deviations of the population sample lead to larger standard errors and larger confidence intervals

We can express various commonly used confidence levels and the corresponding standard errors from mean in Table 1.2.

We now explain the 95% confidence interval using an example for a network experiment.

- Let X1 be the estimate of a performance metric of interest, say end-to-end packet delay in a network simulation experiment.

- Repeat the simulation M times by generating a set of random data and compute the end-to-end packet delay for sufficiently large M. Let these be X1, X2, ... XM.

Table 1.2: Confidence level and the standard error from mean

Confidence level	Standard error from mean
99%	2.58
95%	1.96
90%	1.64
80%	1.28

- Which of $X1, \ldots XM$ is the actual packet delay?

- We first compute the mean of these sample means.

$$\bar{X} = \frac{\sum_{j=1}^{M} X_j}{M}$$

Similarly, we can compute $\sigma_{\bar{x}}$. Now, the confidence interval is $\sigma_{\bar{x}}$.

Example 1.2: A survey was carried out to determine the average annual family income in a city. A sample of 100 families in the city yielded a mean of Rs. $10,000$ and a standard deviation of 1000. Obtain a 95% confidence interval for the true average family income in the city.
Solution: The 95% confidence interval is given by
$$\bar{x} \pm 1.96 \frac{\sigma}{\sqrt{n}}$$
$$= 1000 \pm 1.96 \frac{1000}{\sqrt{100}}$$
$$= 10,000 \pm 196$$
$$= (9,804, 10,196) \ \square$$

Example 1.3: In a network simulation experiment, 100 readings were taken. The following results were obtained. The mean delay is $\bar{x} = 120$ and the standard error is $\sigma_{\bar{x}} = 30$. Obtain the 99% confidence interval for the average delay.
Solution: The 99% confidence interval is
$$\bar{x} - 2.58 \frac{s}{\sqrt{n}}, \bar{x} + 2.58 \frac{s}{\sqrt{n}}$$
That is, the 99% confidence interval is given by
$$(1202.58(30/\sqrt{100}), \ 120 + 2.58(30/\sqrt{100}))$$
$$(1207.74, \ 120 + 7.74) = (112.26, \ 127.74) \ \square$$

Example 1.4: In 20 runs of a network simulation experiment, the mean percentage of packets lost was 9.4. In 18 runs of the 20, the results were found to be accurate within a bound of 1.4 times the mean.
a) What is the margin of error?
b) What is the confidence interval?
c) What is the confidence level in the results expressed in a percentage?
Solution:
a) 1.4 (results accurate within 1.4).

b) $9.4 - 1.4 = 8$, $9.4 + 1.4 = 10.8$. Therefore the confidence interval is $(8, 10.8)$.

c) 18 out of 20 is a 90% confidence level, $\frac{18}{20} = \frac{x}{100}$. □

Example 1.5: With a sample size of 800 and a standard deviation of 4.3, what is the 90% confidence interval if the sample mean is 4.5?
Solution:
$$4.5 \pm 1.645 \frac{4.3}{\sqrt{800}} \quad or \quad 0.25$$
Therefore the 90% confidence interval is
$$(4.50.25 = 4.25, 4.5 + 0.25 = 4.75) = (4.25, 4.75) \square$$

1.7.3 A brief history of computer-aided simulation

Before the advent of computers, a popular technique that was deployed to evaluate the performance of a system was solving a set of mathematical equations modeling the system. However, solving a set of mathematical equations modeling a nontrivial system was complex. As a result, it was difficult to estimate the performance of nontrivial systems, and this approach did not become very popular. Later, with the availability of computers, iterative solutions requiring a large number of computations per iteration could be easily used, and therefore these techniques gained prominence. As computers became more powerful over time, more and more sophisticated computer-aided simulation techniques could be used. We now provide a classification of different computer-aided simulation techniques that were developed.

1950–1960: With the availability of computers, programs based on the Monte Carlo method could be written to solve certain complex problems. An early use of Monte Carlo simulation was for the study of neutron diffusion that arose in the design of the hydrogen bomb, and this problem was known to be analytically intractable (Cooper 1988).

1960–1970: In the early 1960s, simulation programs were mainly written using general purpose programming languages such as FORTRAN. In the 1960s, specialized simulation languages such as SIMCRIPT and GPSS were developed. These languages provided several primitives for basic simulation experiments such as generation of random numbers with different distributions, event handling, keeping the simulation time, basic queueing disciplines, etc. This resulted in widespread use of simulation in areas such as manufacturing and finance.

1970–1980: This was the era of mainframe computers. These were time-shared and multiuser computers, but user interactions were severely limited. In 1967, the programming language SIMULA, primarily focused on programming simulation experiments, was developed. It had the concept of class and therefore is regarded as the first object-oriented programming language.

1980s– the present: Powerful PCs became available at the start of this period. This led to the availability of software packages supporting simulation model creation in a user-friendly graphical environment. Animation gradually became a standard tool in simulation modeling software. Graphical animation lets one visually verify whether the topology created is as per requirement and the data flows appear intuitive. Most of the available software packages included modules for statistical analysis of input and output data. Simulation carried over the internet, called web-based simulation, is a recent trend in the field. Web-based simulations can deploy server-side simulation or client-side simulation. In server-side simulation, the simulation program resides on a server and the user fixes the simulation parameters from a client. The simulation run and graphics animations are computed on the server side and shown on the client browser through Java script or flash programming. On the other hand, in client-side simulations, the simulation program is downloaded and executed on the client browser in the form of a Java applet, flash, or a PHP program.

Distributed simulation: Large network topologies and complex data characteristics can overwhelm the host computer on which the simulation experiment runs. For such situations, distributed simulations are necessary. In this approach, the full simulation topology is divided into N submodels, and each submodel is independently simulated on a different host running the simulation software, and of course the intermediate results are exchanged among different hosts using message passing to generate the overall simulation statistics.

1.7.4 Simulation versus other evaluation techniques

In the absence of availability of the actual system, the performance of the system can be evaluated using a variety of techniques. These techniques are schematically depicted in Figure 1.5. All performance evaluation techniques, in the absence of the real system, involve the use of some model of the system. Such a model can be either a physical model of the system such as a prototype or an emulation test bed, or a mathematical model. A mathematical model can be evaluated through either a closed-form solution or a simulation technique. We briefly compare the relative accuracy of the results obtained by using these techniques in the following. In simulation experiments, a model of the system and the environment is given to a simulation engine, and the simulation engine tries to recreate the behavior of the given system for the given environment. Of course, the behavior created is usually very different from how the real system exhibits its behavior. On the other hand, in emulation, the created behavior is much closer to the behavior of the given system.

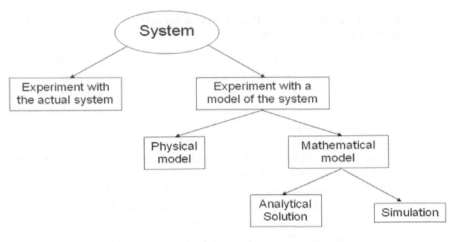

Figure 1.5: Approaches to system evaluation

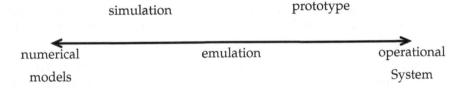

Figure 1.6: Accuracy of results

In other words, it is a truer replication of the system whose behavior is being emulated. For example, a network switch may be configured to emulate a hub. Therefore, the accuracy of a performance evaluation result using a prototype or emulation is much higher as compared to a simulation. This aspect is schematically depicted in Figure 1.6. As can be observed from Figure 1.6, performance evaluation of complex systems through analytical solutions to mathematical models such as queueing models typically yields much lower accuracy than simulation results.

1.7.5 Writing your simulation program versus using a simulation tool

It is usually difficult to write your own simulation program from scratch as compared to using a simulation tool for evaluating a system. The reason is that a simulation program is usually an event-based program where an event may cause further events to arise. This calls for experience in writing event-based programs. Further, the events can arise concurrently, and this makes programming even harder. A simulation tool such as NS2 for network simulation transparently carries out all the steps of the simulation and provides an easy to use interface for the user. It usually provides graphical output, animation capability, support for traffic generation obeying any required probability distribution, and a custom language that helps one to easily create network components such as nodes, routers, etc. In addition, several established protocols are builtin and can be used along with a protocol that one is trying to simulate.

1.7.6   Basic simulation terminology

In the following, we discuss terminology that is commonly used in the network simulation domain.

Random variable: A random variable models a quantity whose value is subject to variations due to chance.

Random variate: A random variate is an artificially generated random variable.

Simulated time: This is an internal variable of the simulation program that keeps track of simulated time.

State variables: State variables are the variables whose values define the current state of a system. For example, the variable length that keeps track of the jobs in the queue in a server system is a state variable. If the length is either zero or a small number, then we say that the system is underloaded. If the length is determined to be a large number, then we say that the servers are overloaded.

Event: An event occurs at a point in time and may affect the system state. An event is internally represented by a data structure, and it stores the event time and the actions to be taken. At the precise time point stored in the event data structure, a simulation program must take certain actions such as changing the state and causing generation of future events. A few examples of events are the arrival of a job, the beginning of a new execution, the departure of a job, etc.

1.8   Discrete-Event Simulation

Discrete-event simulation (DES) is a widely used simulation technique. The term "discrete event" refers to the fact that the state of the system changes only at discrete times, rather than continuously. Events occur at discrete time instances. A typical example is an application involving a server serving queued-up clients. As an example of a DES, consider the following. Suppose packets are arriving and getting queued up at a router. The number of packets in the queue increases only when a packet arrives and decreases only when the processing of the packet at the router is complete and the router forwards it to one of its ports for transmission, both of which occur only at discrete times.

Central to the operation of DES is the maintenance of an event list. An event list is a list of events that have already occurred. Each event contains a time stamp, indicating when the event occurs and the action that needs to take place once the event occurs. At any point of time during simulation, the earliest event that has occurred is handled next. The event list is usually maintained in ascending order of time stamps. The main loop of the simulation sets up an iteration. Each iteration takes out the earliest event from the event list, updates the simulated time to reflect the occurrence of that event, and may result in the creation of new events that are queued up in the event queue. The following code illustrates the main steps that are needed to be carried out in a discrete-event simulation program.

Listing 1.1: Discrete Event Simulation program

```
1  main(){
2    while(eventList not empty){
3      nextEvent=eventList.head();
4      eventList.deleteHead();
5      simulationTime=nextEvent.time;
6      processEvent(nextEvent);
7    }
8    . . .
9  }
```

A typical simulation experiment
Figure 1.7 shows the schematic of a typical simulation experiment to simulate the working of a router. Packets arrive with average interrarrival time: $1/\lambda$ at the router. The average execution times are $1/\mu1$ and $1/\mu2$ for the two outgoing links. The arriving packet joins link i with probability ϕ_i. The state of the system is given by the sizes of the queues at the different routers. The state of the system changes due to the arrival of packets and the completion of routing at the different routers.

Transient behavior
Simulation outputs that depend on the initial conditions are called the transient characteristics of the system. The transient characteristics cause the output to change when the initial conditions change. The transient characteristics are usually prominent during the "early" part of simulation. The later part of simulation is less dependent on initial conditions. We illustrate the potential impact that the transient conditions can have on the output using a hypothetical simulation output. Figure 1.8(a) shows a histogram of the delay experienced by a packet, given an initially empty queue at a router. Figure 1.8(b) shows a histogram of the delay experienced by a router, given

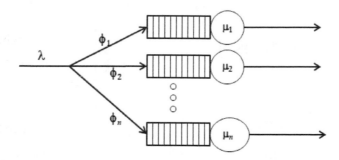

Figure 1.7: Schematic of simulation of a router

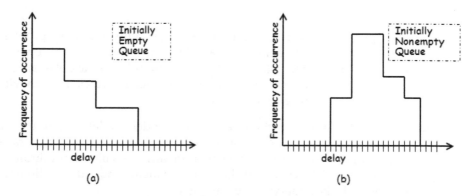

Figure 1.8: Simulation output showing effect of transient condition

nonempty initial conditions.

1.8.1 Model validation techniques

The correctness of simulation results depend to a large extent on whether the model closely corresponds to the real system. Model validation techniques are used to check if the constructed model is a true replica of the system being evaluated. There are a few well-established approaches to accomplish this. We briefly discuss these in the following.

Measurements on real systems Validation of a model by comparing the behavior of the model with that of the actual system is possibly the most reliable way of validating the simulation results. However, this approach is often not feasible because the system may not exist or it may be too expensive to measure and experiment on the actual system.

Continuity tests Continuity implies that a slight change in input should yield only a slight change in output. However, kinks and spikes in

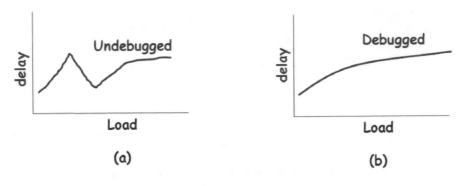

Figure 1.9: Continuity test

the input versus output plot indicate discontinuity and likely problem in the simulation experiment. Examples of discontinuity are shown in Figure 1.9. In Figure 1.9(a), several discontinuities in the delay are noticeable with a change in load. After the model is corrected, a behavior such as that in Figure 1.9(b) may be obtained.

Degeneracy tests Another technique used for model validation is to try out the extremes of a parameter such as the highest and lowest data rates and examine the simulation results for any possible discontinuities. Sometimes, mistakes in simulation experiments show up as discontinuities in the boundary of the parameters.

Consistency tests These tests try to ensure that similar inputs produce similar outputs. For example, if, for a certain network simulation experiment, two data sources produce traffic at the rate of 50 pkts/sec each, then one would expect somewhat similar behavior when there is only one source that produces 100 pkts/sec traffic.

1.9   Basics of Network Simulation

Most network simulations are based on discrete-event simulation techniques. In these, state changes occur at discrete points in simulation time due to events such as message generation, packet arrival, packet departure, etc. On the occurrence of each event, certain processing must be done and this is taken in the time stamp order of event arrival. At present, several packages are available for carrying out network simulations. A few of the popular ones are NS2, Opnet, and GloMoSim.

The first step in carrying out a network simulation experiment is to create a model of the network topology. This can be achieved by defining a set of nodes and their interconnections. The different aspects of network topology that are normally specified are as follows.

Network routers: The different aspects specified for network routers include the number of network interfaces, the characteristics of the physical media used by the different interfaces, the type of queueing discipline deployed at each of the output interfaces, the type of routing protocol used by each interface of the router, and the input and output queueing disciplines deployed. The last parameter is usually specified for high-performance routers, for which both the ingress queue (input queue) and the egress queue (output queue) may have to be specified.

Communication links: The different aspects specified for a communication link include the transmission rate of the link (for example, 100 Mbps), the propagation delay of the medium, and whether the link is full duplex or half duplex.

End systems: The different aspects that need to be specified for an end system usually include the protocols (e.g., IPv4, TCP, UDP, etc.) supported, the delay experienced while packets progress up and down the protocol stack, and the action that the end-system performs in response to illegal or unexpected packets.

The other aspects typically specified for a simulation experiment include the traffic characteristics. For example, for simulating a web application, the probability distribution of the request to the web server, the size of the response of the server, and the amount of processing time required by the server need to be specified.

1.10   Introduction to NS2

NS2 is a free and open-source discrete-event simulator. A major attraction of using NS2 is that it has several common protocols built in, and so are different types of traffic generators, link and node characterizations, etc. Some of the network protocols that are built into NS2 include TCP/IP suite and various other protocols to perform experiments in both wired and wireless mediums. NS2 uses a simple programming model by having a single thread of control. It simplifies many things since there are no locking or race conditions to worry about.

A brief history of NS2
The NS simulator grew out of the REAL simulator developed by Srinivasan Keshav in 1989. Later it was extended by the Network Research Group at the Lawrence Berkeley National Laboratory by Floyd and McCanne in 1995. NS2 later grew out of this work by the VINT project (Virtual InterNetwork Testbed) at the Lawrence Berkeley National Laboratory. Of late, NS3 has been

introduced to address some of the shortcomings of NS2. The core of the NS2 simulation engine is implemented using C++. However, to set up a simulation experiment, the topology needs to be described using oTCL, a scripting language. The reason behind this is that, since the simulator was designed in the 1990s, desktops were rather slow. The topology had to be specified multiple times, and the errors in topology specification had to be corrected many times. The compilation of C++ programs was very time consuming, whereas an interpreted language such as oTCL required interpretation of the changed line. However, with the availability of powerful hardware, compilation time is hardly an issue. Therefore NS3 has been implemented in C++ entirely. NS2 comes with a package called NAM (Network Animator), a Tcl-based animation system that produces a visual representation of the specified network. On the other hand, NS3 employs a package known as PyViz, which is a python-based real-time visualization package.

1.11   Common Mistakes in Simulation

In the following, we discuss a few common mistakes in carrying out simulation experiments. One needs to be aware of these mistakes and consciously try to avoid committing them.

- **Simulation duration too short:** In an attempt to save time, one may be tempted to run the simulation experiments in a short time. However, this may not only lead to inaccurate results, but could also make the results more dependent upon initial conditions. The correct length of simulation runs depends upon the accuracy desired (confidence intervals). A considerable run of the simulation experiment is akin to a larger sample size and makes the confidence interval shorter and the results more accurate.

- **Use of poor random number generators and seeds:** This problem often arises when one writes a simulator rather using a professional simulation tool. The generated numbers may not be random enough; as a result, different results are correlated and even identical, and the simulation results become inaccurate.

- **Other common problems:** Other common problems include not taking care to validate the model and failing to notice surprising results such as kinks and discontinuities in the output.

Exercises

1) What is meant by a data network? Differentiate between a data network and a plain old telephone (POTS) network.

2) What is ISDN? What are its advantages and disadvantages?

3) Which one of the following mobile wireless networks can be considered a data network?

    a. 1G

    b. 2G

    c. 3G

    d. CDMA

4) Briefly explain how TCP/IP has been incorporated into the Unix operating system.

5) Name two uses each of TCP and UDP.

6) Distinguish between a data network and a voice network.

7) What is the difference between continuous-time and discrete-event simulation? Give one example of each. Point out the relative advantages of continuous-event and discrete time simulations.

8) How can simulation results be validated?

9) Other than simulation, what are other approaches available to evaluate a network protocol? How do these compare to simulation?

10) What do we mean by a confidence interval in a simulation result? Explain your answer with an example. In which situations is it advantageous to use confidence intervals? What problems can arise if confidence intervals are not used and rather point estimates are used?

11) Briefly explain the advantages and disadvantages of simulation as compared to (a) experimenting on the actual system, (b) solving an exact mathematical model of the system, (c) solving a statistical model such as a queueing model of the system.

12) What is a socket in the context of network-based programming? What do you understand by socket programming?

13) What is the difference between a data network and a voice network? What do you understand by integration of voice and data networks? Briefly explain how this is achieved.

14) In achieving integrated services, is data transmission over voice networks or voice transmission over voice networks more successful?

15) Give at least three advantages of using a network simulator such as NS2 as compared to writing your own network simulator program using languages such as C or Java.

16) Give one important advantage of circuit-switched communication over packet-switched communication.

17) Give one important advantage of packet-switched communication over circuit-switched communication.

18) In the absence of the real system, what are the different techniques that are available to evaluate the performance of the system?

19) Discuss the pros and cons of evaluating the performance of a networking system by simulating versus carrying out experiments on the actual system.

20) Discuss the pros and cons of writing your own program to simulate a network protocol such as TCP, versus using a standard network simulation software application such as NS2.

21) Suppose you wish to write your own discrete-event simulation program. Give the main steps that must be programmed in the form of a pseudocode.

22) Explain the difference among the following types of simulation techniques:

   a) Static versus dynamic

   b) Stochastic versus deterministic

   c) Discrete event versus continuous time

23) What are the main problems that could be encountered while transmitting digital data over a voice network? What are the ways that they can be overcome?

24) What are the main problems that could be encountered while transmitting analog voice signals over a data network? What are the ways that they can be overcome?

25) How can simulation results be validated?

26) What are some of the common mistakes that are committed in simulation experiments and against which you need to take special precautions?

27) What are the main steps carried out by a discrete-event simulation program? Indicate these steps in the form of a pseudocode.

28) Compute the 95% confidence interval if you notice that packets take 12 msec to reach the destination and the standard error is 2 msec.

29) What are the components of a queueing system? Queueing theory is applicable to the performance evaluation of which systems? Give examples.

30) What is an M/M/1 queue? To which type of systems is it applicable?

31) Messages arrive at a switching center at an average rate of 100 messages per minute. The message length is exponentially distributed with an average length of 10 characters. The arrival pattern follows a Poisson process. The bandwidth of the outgoing line is 80 characters per second. Calculate the following statistical measures of system performance:

    a. Average number of messages in switch

    b. Average number of messages in the queue to be transmitted

    c. Average delay that a message encounters at the switch

    d. Average time that a message waits for transmission

32) What are some of the performance measures of a system that can be determined from the use of queueing theory?

Bibliography

[1] Banks, J., Carson, S. J., Nelson, B. N., and Nicol, D. M. 2005. *Discrete-Event System Simulation* (4th ed.). New Jersey: Prentice-Hall, Inc.

[2] Cooper, N.C., ed. 1988. *From Cardinals to Chaos: Reflections on the Life and Legacy of Stanislaw Ulam.* Cambridge: Cambridge University Press.

[3] Gross, D., Shortle, J. F., Thompson, J. M., Harris, C. M. 2008. *Fundamentals of Queueing Theory.* New York: Wiley.

CHAPTER 2

NETWORK PROTOCOLS

Computer communication is a complex task. The complexity lies not only in making the communication between various machines reliable and efficient, but also supporting various media types, computers with very different hardware and speed, and the fact that various network components may be manufactured by different vendors and may need to be upgraded frequently considering that networking is a dynamic field. This complexity has been attacked by deploying two main technologies in the solution. These are standardization and layering.

Computer communication is achieved through a set of protocols. The layering and standardization of protocols played a key role in the rapid acceptance and growth of computer networks. Standardization is vital to the satisfactory operation of computer networks, as different parts of a network often operate using hardware and software manufactured by different vendors.

The International Organization for Standardization (ISO) formed a subcommittee named OSI (Open Systems Interconnection) to oversee the standardization work. ISO/OSI proposed a seven-layer protocol model for computer communication in 1979. A cornerstone of the ISO/OSI model is the organization of various networking protocols into well-defined layers. Layering helps to overcome the complexity of network operations and facilitates rapid evolution of network protocols. In the ISO/OSI seven-layer structure, each layer provides certain well-defined services to its upper layer. The ISO/OSI protocol now serves as more of a theoretical reference, whereas TCP/IP can be considered a practical implementation of the ISO/OSI protocol suite.

2.1   TCP/IP Protocol Suite

The TCP/IP protocol suite was developed by DARPA in 1969 to provide seamless communication services across an internetwork consisting of a large number of different networks. The TCP/IP protocol suite is a collection

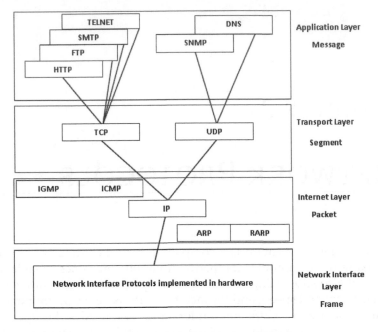

Figure 2.1: TCP/IP Protocol Stack

of a large number of protocols. The protocol suite is named after the two important protocols of the protocol suite: Transmission Control Protocol (TCP) and Internet Protocol (IP). A few important protocols in the TCP/IP suite and the specific protocol layer at which they operate are depicted in Figure 2.1. The diagram also depicts the specific lower-layer protocols that are invoked by a protocol. As shown in Figure 2.1, the TCP/IP protocol stack consists of four layers of protocols. The four layers are application layer, transport layer, internet layer, and network interface layer. The TCP/IP does not define any specific protocol for the network interface layer, but allows any of the standard protocols, Ethernet, HDLC, and Token ring, to be used at this layer. For this reason, in Figure 2.1 we do not show any specific protocol for the network interface layer. Note that the network interface layer is typically implemented by hardware either directly as a set of chips on the motherboard or as a plug-in card called the Network Interface Card (NIC). The MAC address is defined by this hardware. The TCP/IP and the application layer protocols are implemented by software.

Application programmers and end users are mainly concerned with application layer protocols. Application layer protocols, in turn, make use of the services provided by the lower layer protocols. An application layer protocol sending a message to another application (that may possibly be running on a different hosts either in the same local network or in a remote network) makes use of a transport layer protocol and passes to it the message to be transmitted. The specific transport layer protocol converts the message into small parts and attaches certain information to it. The transport layer protocol first converts a message into segments and passes these segments to the Internet layer protocol (IP). The IP layer protocol attaches certain information to the segments, such as the destination host address, to form packets. We can say that a TCP segment is carried in one or more IP packets. The IP passes on the packet to the network interface layer protocol, which in turn converts it into frames by adding certain additional information, such as check sum, and transmits it on the network.

The reverse operation takes place when a frame arrives at a host. The network interface layer protocol removes the information added by the corresponding network interface layer protocol at the sender end and passes on the packet to the IP layer. The IP layer protocol at the destination removes the information added by the IP layer at the sender's end and gets back the segments and passes these to the transport layer protocol. The transport layer protocol at the receiver strips the information added by the transport layer protocol at the sender, reconstructs the message, and sends it to the application layer.

Note that the basic data units that different protocol layers deal with are different. For example, the application layer deals with messages, the transport layer deals with segments, the internet layer with packets, and the data link layer with frames. The terminology used to refer to the unit of data transfer at the different layers is shown in Figure 2.1.

Over the last two decades, the Internet has seen almost exponential growth and now Internet applications have become ubiquitous. Internet-based applications are developed predominantly by using the client-server paradigm. In a typical Internet-based application, the application consists of a server part and a client part. The server is an application providing certain services and a client typically runs on a web browser and are primarily the requester of services. TCP has now become the de facto transport layer protocol for client-server communications.

2.2   Terminology

In the following, we briefly discuss a few of the protocols and the terminology associated with the TCP/IP protocol suite.

TCP (Transmission Control Protocol): On the sending side, TCP is responsible for breaking a message into small parts. It adds sequence numbers and certain other information which after these become known as segments. TCP passes the segments to the lower layer protocol for transmission over the network. At the receiver's end, TCP assembles the segments when they arrive and reconstructs the message. TCP is a reliable protocol. When a packet is lost or corrupted during transmission, TCP detects it and requests the sender to retransmit. Thus retransmission is used as the primary mechanism by TCP for reliable data delivery to the destination. The IP layer achieves transmission of a packet to the receiver machine. However, there may be many applications running on the receiver machine. TCP makes a segment reach the required application based on the port number.

UDP (User Datagram Protocol): For many applications, reliable transmission is not required. These applications simply ignore a delayed or lost packet at the receiver end. For these applications, the overhead of retransmission is avoided as the retransmitted packages would be ignored. An example of such an application is Ping.

IP (Internet Protocol): At the host machine of an application sending a message, IP is responsible for constructing packets (also called datagrams) from the segments it receives from the transport layer protocol by adding the destination host address and passing these on to the lower layer protocol for transmitting. On the receiver's side, it deconstructs the segment and passes on the segments to the transport layer protocol.

HTTP (Hyper Text Transfer Protocol): HTTP is used for communications between a web server and the client-side application running on a web browser. Typically the web browser initiates an HTTP request to the web server. Immediately after this, the client disconnects from the server and waits for a response. The server processes the request, possibly by gathering an HTML page from its database, and reestablishes the connection with the client to send the HTML page as its response. HTTP is a stateless protocol, because both client and server do not remember the last activity.

SMTP (Simple Mail Transfer Protocol): SMTP is used for sending and receiving e-mails by a mail client. SMTP is used with protocols such as POP3 (Post Office Protocol) for client-side storage.

MIME (Multipurpose Internet Mail Extensions): The MIME protocol lets SMTP encode multimedia files such as voice, picture, and binary data in e-mails and transmit them across TCP/IP networks. SMTP has been designed to handle only text contents in e-mails. MIME helps e-mails to include non-text contents such as picture, voice, and binary data files through definition of binary data suitable for encoding in ASCII text format.

FTP (File Transfer Protocol): FTP is used to transfer files between computers. Ftp opens two TCP connections between the client machine and the server machine for transfer of data and control commands, respectively.

SNMP (Simple Network Management Protocol): SNMP is used for administration and management of computer networks. The network manager uses tools based on this protocol to monitor network performance.

ICMP (Internet Control Message Protocol): ICMP runs on all hosts and routers and is mainly used for reporting errors, such as a nonreachable host.

ARP (Address Resolution Protocol): ARP is used by IP to find the hardware address (also called the physical address) of a computer based on its IP address. The hardware (physical) address is stored in the ROM (Read Only Memory) of the computer's network interface card. It is also known as the

MAC (Media Access Control) address and as an Ethernet hardware address (EHA).

RARP (Reverse Address Resolution Protocol): RARP is used by IP to find the IP address based on the physical (MAC) address of a computer.

BOOTP (Boot Protocol): BOOTP is used for booting (starting) a diskless computer over a network. Since a diskless computer does not store an operating system program in its permanent memory, BOOTP helps to download and boot over a network, using the operating system files stored on a server located in the network (discussed in Section 4.7).

Routers: A router is responsible for routing the packets that it receives to their destinations based on their IP addresses, possibly via other routers.

DNS: This stands for Domain Name System (or Service or Server). It is a software service available on the Internet that is responsible for translating domain names into IP addresses. We use domain names while accessing any web site since these are alphabetic character strings that are much easier to remember compared to the conventional IP address specification using dot-separated numerical values. Of course, when we specify a web site (URL) using its domain name, a DNS service hosted on the Internet translates the domain name into the corresponding IP address, since, after all, the Internet works using IP addresses. For example, the domain name

www.iitkgp.ernet.in

might be translated by the DNS to 144.16.192.245.

IP Addresses: Each computer must have an IP address before it can be meaningfully connected to the Internet. A packet is routed to its destination based on its IP address.

IGMP (Internet Group Management Protocol): IGMP is used by hosts to exchange information with their local routers to set up multicast groups. The set up of multicast groups allows efficient communication, especially for video streams and certain gaming applications. Routers also use IGMP to check whether members of a known group are active or not.

2.3 Architecture of TCP/IP

TCP/IP consists of four layers, as shown Figure 2.2. These layers are application layer the transport layer, Internet layer, and network access layer.

The functionalities of each of these layers are discussed below.

Application layer: The protocols at this layer are used by applications to establish communication with other applications which may possibly be running on separate hosts. Examples of application layer protocols are http, ftp, and telnet.

Transport layer: This provides reliable end-to-end data transfer services. The term end-to-end means that the end points of a communication link are applications or processes. Therefore sometimes protocols at this layer are also

| Application Layer |
| Transport Layer |
| Internet Layer |
| Network Access Layer |

Figure 2.2: TCP/IP layers

referred to as host-to-host protocols. Remember that there can be several applications or processes running on a host. Thus, to identify the end point, not only does the computer need to be identified, but the exact process or application that would receive the message needs to be identified. This is efficiently accomplished by using the concept of a port number. An application or a process specifies a port number on which it would receive a message. Once a message reaches a host, it is demultiplexed using the port number at the transport layer for delivery to the appropriate application. The transport layer provides its services by making use of the services of its lower layer protocols. This layer includes both connection-oriented (TCP) and connectionless (UDP) protocols.

Internet layer: The Internet layer packs data into data packets that are technically known as IP datagrams. Each IP datagram contains the source and destination address (also called the IP address) information that is used to forward the datagrams between hosts and across networks. The Internet layer is also responsible for routing of IP datagrams. In a nutshell, this layer manages addressing of packets and delivery of packets between networks using IP addresses. The main protocols included at the Internet layer are IP (Internet Protocol), ICMP (Internet Control Message Protocol), ARP (Address Resolution Protocol), RARP (Reverse Address Resolution Protocol), and IGMP (Internet Group Management Protocol).

A host can run many client and server applications. Therefore these different applications can send/receive data concurrently and independently. As a result, data sent by different applications needs to be multiplexed together before they are sent on the network. Similarly, TCP receives segments that may correspond to different applications running on a host. Therefore, on receiving a segment from its lower layer, TCP has to decide which application is the recipient. This is called demultiplexing. TCP performs multiplexing and demultiplexing by using port numbers.

Network access layer: The functions of this protocol layer include encoding data and transmitting at the signaling determined by the physical layer. It also provides error detection and packet framing functionalities. As we discuss in Section 2.6, the functionalities of this layer actually consist of the functionalities of the two lowermost layers of the ISO/OSI protocol suite, namely, data link and physical layers. The data link layer protocols help deliver data packets by making use of physical layer protocols. A few popular data link layer protocols are Ethernet, Token Ring, FDDI, and X.25. Ethernet is possibly the most common data link layer protocol. The physical layer defines how data is physically sent through the network, including how bits are electrically or optically signaled by hardware devices that interface with a network medium, such as a coaxial cable, optical fiber, or a twisted pair of copper wires.

2.4   Overview of the Operation of TCP

When a client-server application runs on hosts that are far apart, data transmission between the client and the server may span multiple networks. These networks are called sub-networks. For data routing, Internet protocol (IP) requires that each host in the network have a unique address. Identification of the host is not enough for data deliveries; the packets must be forwarded to the exact application (or a process in an application) requiring the packet. Within each host, every process is identified by a port number, based on which the transport (TCP) can deliver data/information to each relevant process.

Usually a message in the form of a block of data is passed to TCP by the sending application. TCP breaks it into many small parts and attaches certain control information (called the TCP header) to each small part. Each small part of the data along with the TCP header is called a segment (see Figure 2.3).

The TCP header includes the destination port, checksum, and sequence number.

For a more detailed treatment of TCP operation, the reader is referred to [13].

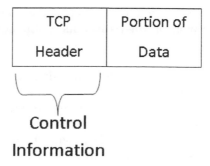

Figure 2.3: The structure of a TCP segment

IP datagram

An IP packet is also called a datagram. A datagram is of variable length which can be up to 65,536 bytes. It has two fields, namely, header and data. The structure of an IP datagram is schematically shown in Figure 2.4. In the following, we discuss some of the important fields of an IP datagram.

Version (Ver): The IP version number is defined in this field, e.g., IPV4 or IPV6. *Header length (Hlen):* This defines the header length as multiples of four bytes. *Service type:* This has bits that define the priority of the datagram itself. *Total length:* This field is allotted 16 bits to define the length of an IP datagram. *Identification:* This is mainly used to identify fragmentation that belongs to different networks; 16 bits is allotted for this job. *Flags:* This deals with fragmentation of the data. *Fragmentation offset:* This is a pointer to the offset of the data in the original datagram. *Time to live:* This field is used to define the total number of hops that a datagram has to travel before the discarding operation. *Protocol:* This field has bits. It defines which upper layer protocol data is encapsulated at that time for example, TCP or UDP or ICMP, etc. *Header checksum:* This has 16 bit fields to check the integrity of the packets. *Source address:* This is a 4 byte (4*8 = 32) Internet address to define the original source. *Destination address:* This is a 4 byte (4*8 = 32) Internet address to identify the destination of the datagram.

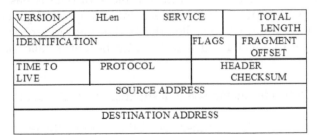

Figure 2.4: Datagram structure

IP datagrams are sent to the link layer and become ready for transmission to the first subnetwork in the path to its destination. IP datagrams are also called packets.

Port address

In a client-server application, often the client and server programs are located on different host machines. The client program usually uses a temporary port number and the server program uses a well-known (or permanent) port number. These port numbers are used for identification of the application. A few well-known ports used by some of the popular TCP/IP protocols and by user applications are given in Table 2.1.

Data Encapsulation

When the TCP segments are handed over to IP layer, it appends an IP header containing relevant control information. A segment after the additional con-

Table 2.1: A few commonly used well-known port numbers

Protocol	Port
TELNET	23
SMTP	25
RPC	111
DNS	53

trol data has been added is called an IP datagram (see Figure 2.4). The network access layer also appends its own header and now it becomes known as a frame/packet. The packet header includes important information such as facilities requests (such as priority) and the destination subnetwork address.

2.5   Application Layer Protocols of TCP

The application layer protocols of the TCP/IP suite are shown in Figure 2.5. The following are three important application layer protocols.

Simple Mail Transfer Protocol (SMTP): This provides an electronic mail function that is used for transferring messages between different hosts. Originally, SMTP could handle text messages only. MIME helps transmit multimedia data within an e-mail by encoding the binary multimedia data in ASCII format.

File Transfer Protocol (FTP): FTP is mainly used for transferring files from one host to another based on a user command. Ftp allows both binary and text file transfers. Each ftp connection opens two TCP connections, one for data transfer and the other for transfer of control commands such as put, get, etc.

TELNET: This application layer protocol lets users use a remote log-on facility, using which a user can log on to a remote system. Both ftp and TELNET make use of the TCP layer. TCP forwards these data over the network by invoking the IP layer, and the IP layer in turn invokes the link layer protocol. A problem with this type of transmission is that it becomes easy to sniff (secretly hear) the data by using publicly available TCP sniffer programs such as TCPdump. Due to this, at present most users and applications use sftp (secure ftp) and ssh (secure shell) protocols. These essentially serve to encrypt the data before passing these on to the TCP layer. These also perform decryption after receiving data.

2.6   TCP/IP versus the ISO/OSI Model

It is important to realize the difference between the TCP/IP protocol suite and the ISO/OSI model. As already mentioned, TCP/IP was developed by DARPA (Defence Advanced Research Program of the US government) in 1969. The aim was to integrate all computers efficiently. Thus TCP/IP

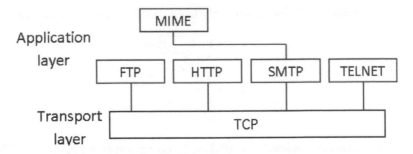

Figure 2.5: Application layer protocols in TCP/IP

networking predates ISO/OSI, which was formulated in 1976. Naturally, TCP/IP makes no reference to the ISO/OSI model and splits the network functionalities into layers quite differently.

A rough correspondence between TCP/IP layers and ISO/OSI layers is shown in Figure 2.6. Observe that the Internet layer roughly corresponds to the network layer of the ISO/OSI model. The network access layer encompasses the data link and physical layers. The TCP/IP protocol suite does not define specific data link layer protocols to be used and can work on any data link protocol such as Token Ring and Ethernet.

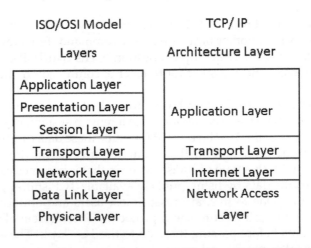

Figure 2.6: Comparison of TCP/IP and ISO/OSI

2.7 Adaptation of the TCP Window

TCP primarily deploys a flow control technique to control congestion in a network. Traffic congestion occurs when the rate at which data is injected by a host into the network exceeds the rate at which data can be delivered to the network. A flow control technique helps adapt the rate of data transmission by the TCP at the sending host end. The flow control technique helps to prevent build up of congestion in the network and at the same time helps to prevent buffer overrun at slow receivers.

If data transmissions occur at a much faster rate than what the network infrastructure can comfortably support, then data packets get built up at the routers. When the buffers at routers start to overflow, packets start to get lost. Additionally, if data transmissions by a sender take place at a much faster rate than what a slower receiver can handle, then the receiver's buffer starts to get flooded and packets get lost. TCP handles both these causes of packet loss by reducing the rate at which data is transmitted at the sender's end. Thus a receiver uses the flow control mechanism to restrict how fast a sender can transmit. However, to provide acceptably fast data transmission service, once congestion disappears, the transmission rate at the sender's end needs to be increased to a suitable value. Thus we can say that a flow control technique helps TCP dynamically adjust the transmission rate at the sender's end, reducing the rate as congestion starts to develop and increasing it as congestion starts to disappear. The flow control mechanism deployed by TCP (called the sliding window protocol) is primarily based on the concepts of congestion window and advertised window. We now briefly describe these techniques.

When a sender starts to send data packets, the receiver indicates an advertised window (or receiver window) to the sender while sending acknowledgments. The advertised window is usually set equal to the size of the receive buffer at the receiver. The sender uses the advertised window size obtained from the receiver to determine the maximum amount of data that it can transmit to the receiver without causing buffer overflow at the receiver. In other words, to prevent buffer overflow at the receiver, the data packets transmitted by a sender without having received acknowledgments for them should not exceed the size of the buffer available at the receiver.

For each segment sent, a sending host expects to receive an acknowledgment. A congestion window indicates the maximum number of segments that can be outstanding without the receipt of the corresponding acknowledgment before the TCP at the sender's end pauses transmitting and waits for an acknowledgment to arrive. The TCP at a sender's end pauses if the number of segments for which acknowledgment is outstanding becomes equal to the congestion window. A sender sets the congestion window size to 1 and keeps on increasing it until duplicate acknowledgments are received or the number of outstanding packets equals the advertised window.

Upon receipt of an acknowledgment, TCP detects packet loss using retransmission timeout (RTO) and duplicate acknowledgments. After transmitting a segment, a TCP sender sets the retransmission timer for only one packet. If acknowledgment of the packet is not received before the timer goes off, the packet is assumed to be lost. RTO is dynamically calculated. Timeouts can take too long. Again a TCP sender assumes that a packet loss has occurred if it receives three duplicate acknowledgments consecutively. In TCP, when a receiver does not get a packet it expects in a sequence but gets an out of order packet, it considers that the expected packet might have been lost, and it indicates this to the sender by transmitting an acknowledgment for the last packet that was received in order. Thus three duplicate acknowledgments are also generated if a packet is delivered at least three places beyond its in-sequence location.

In wired networks, packet losses primarily occur on account of congestions encountered in the transmission path. However, in a wireless environment packet losses can also occur due to mobility and channel errors. In wired networks, bit errors are rare. On the other hand, wireless networks are vulnerable to noise. This can cause intermittent bit errors. Further, there can be intermittent disconnections due to fading and also due to obstructions that may be encountered by a mobile host. Further, packets may get lost during hand off. An intermittent disconnection may cause the TCP at the sender's end to time out for an acknowledgment and cause it to retransmit. This would cause unnecessary retransmissions to occur, even though the packet may be buffered at a router. Also various additional causes of packet loss can result in a high rate of packet loss in a mobile wireless network. These would be interpreted by TCP as symptoms of congestion and force it to operate in slow start. This would cause the network to operate inefficiently and result in unacceptably slow data transmission.

When many packets are transmitted to a single receiver and the rate at which these packets are transmitted is higher than the processing rate of the destination host or an intermediate router, the buffers of the router (or the destination, as the case may be) get filled quickly. This results in the affected router or the destination dropping packets. However, a sender realizes that some packets might have been dropped based on acknowledgments not arriving before time out or the receipt of three duplicate acknowledgments. Once a sender realizes that, possibly due to congestion, a packet might have been dropped, it tries to retransmit the missed packet. However, although this mechanism may overcome packet losses, it does not resolve the congestion problem. The congestion problem is resolved through an adaptive transmission control mechanism called flow control.

2.8 Improvement of TCP Performance

TCP was designed for traditional wired networks. If used as it is, a few shortcomings become noticeable. We first review a few relevant aspects of traditional TCP. Then we discuss how TCP has been extended to work efficiently in a mobile environment.

2.8.1   Traditional networks

In wired networks, packet losses are primarily attributable to congestion that is built up in the network. To reduce congestion, TCP invokes congestion control mechanisms. Congestion control is primarily achieved by reducing the transmission window, which in turn results in slower data transfer. The important mechanisms used by TCP for improving (TCP-Reno model) performance are given below.

Slow start

The slow start mechanism is used when a TCP session is started. Instead of starting transmission at a fixed transmission window size, the transmission is started at the lowest window size and then doubled after each successful transmission. The rate of doubling is tied to the rate at which acknowledgments come back. Thus the doubling of window size occurs at every round trip time (RTT). RTT is the time that elapses between a segment being transmitted by a sender and the corresponding acknowledgment being received. If congestion is detected (indicated by duplicate acknowledgments), the transmission window size is reduced to half of its current size and the congestion avoidance process starts. We can say that the rate doubling and reduction is nothing but a binary search technique deployed to determine the "right" transmission window size.

The slow start process begins by the sender setting the transmission window size to 1 and transmitting one segment to the receiver. The sender does not transmit the next segment until it receives an acknowledgment for previous segment. Once the acknowledgment is received, the sender is sure that the congestion window (network capacity) is at least one segment. To determine the exact congestion window size, the sender doubles the transmission window size. It transmits two segments and, after the arrival of two corresponding acknowledgments, it again increases the transmission window size by 2 (to 4) and so on. Increments to the size of the congestion window can be seen to be exponential. A congestion window is doubled every time the acknowledgment arrives promptly. This exponential growth of the congestion window stops at the congestion threshold.

Congestion avoidance

The congestion avoidance algorithm starts where the slow start stops. When the congestion window reaches the congestion threshold level, if an acknowledgment is received, the window size is increased linearly and window size doubling is avoided. From this point, TCP increases its transmission rate linearly by adding one additional packet to its window each transmission time. If congestion is detected at any point, TCP reduces its transmission rate to half the previous value. Thus TCP seesaws its way to the right transmission rate. Clearly, the scheme is less aggressive than the slow start phase, and effects a linear rather than the exponential growth in the slow start process.

Fast retransmit/fast recovery

Usually, a sender initiates a timer after transmitting a packet and sets the time out value RTO. RTO is calculated based on RTT. The sender waits for an acknowledgment from the receiver of a transmitted packet until the timer expires. When the timer expires, it retransmits the packet. However, there exists another situation in which a sender can retransmit a packet. This mechanism is called fast retransmission. In this, the retransmission is not triggered by a timer, but by the receipt of three duplicate copies of an acknowledgment for a packet received from the receiver. Since duplicate acknowledgments also arise when a segment is received out of order, the sender waits for three copies of an acknowledgment for the same packet. This is taken by the sender as the confirmed indication of a missed packet and it starts to retransmit the particular packet. When retransmission occurs, the congestion window size is reduced by half. For example, if the current congestion window size is 4 segments, it is set to 2 segments. Once the lost segment has been retransmitted, TCP tries to maintain the current data transmission rate by not going back to slow start. This is called fast recovery. In fast recovery, the congestion window size is incremented by three since the retransmission occurred after the third duplicate acknowledgment. This is construed as an indication that three packets would have been successfully buffered at the receiver end. Thus, in fast recovery, compensation for the segments that have already been received by the receiver is carried out. If the acknowledgments are received smoothly, this is considered an indication that there is no congestion.

2.8.2   TCP in mobile networks

On the Internet, TCP is the de facto standard transport protocol. It has been remarkably successful in supporting the diverse applications which drive the Internet's popularity. A few such applications are access to information hosted across various web sites, file transfer, and e-mail. The performance of TCP is considered satisfactory for wired networks, but it suffers from serious performance degradation in wireless networks. Wired networks are significantly different from wireless mobile networks. The main differences are much lower bandwidth, bandwidth fluctuations with time and also as a mobile host moves, higher delay, intermittent disconnections, high bit error rate, and poor link reliability. An implication of this difference is that packet losses are no longer due only to network congestion; they may well be due to intermittent link failures, bit error due to noise, or handoffs between two cells. Therefore the traditional TCP assumptions are not valid in mobile (wireless) environments. This leads to poor performance of TCP in mixed wired-wireless environments. Several modifications at the transport layer have been proposed and studied in recent years to deal with the problems. To understand this work, we need to first consider single-hop wireless networks (such as wireless LANs). Based on this, we shall discuss multihop wireless networks (such as mobile ad hoc networks). A few important mechanisms used to improve TCP performance over mobile wireless networks are discussed below.

TCP in Single-Hop Wireless Networks

We first discuss the modifications proposed to TCP to make it effective in single-hop wireless networks.

I-TCP

Indirect TCP (I-TCP) was proposed by Bakre and Badrinath [5]. It segments the connection from the fixed host to the mobile host into two different connections: the wired part and wireless part (see Figure 2.7). A wired connection exists the fixed host (FH) to the base station (BS) and the wireless connection from the BS to the mobile host (MH). Thus the base station maintains two separate TCP connections: one over the fixed network and the other over the wireless link. The wireless link usually has poor quality communication, but it is hidden from the fixed network at the BS. When a packet is sent by the FH to the MH, the packet is received by the BS first and then it transmits it to the MH over the wireless link. If the mobile host moves out of the current BS region, the connection information and responsibilities that are with the current BS are transferred to the new BS.

The advantage of the split connection approach of I-TCP is that it does not need any changes to be made to the standard TCP protocol. By partitioning a TCP connection into two connections in I-TCP, the transmission errors in the wireless part would not propagate into the fixed network, thereby effectively increasing in bandwidth over the fixed network. An important disadvantage of this scheme is that I-TCP does not maintain the semantics of TCP as the FH gets acknowledgment before the packet is delivered at the MH. I-TCP does not maintain the end-to-end semantics of TCP and assumes that the application layer would ensure reliability.

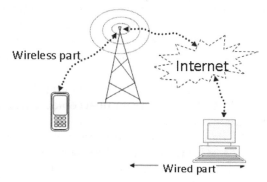

Figure 2.7: A schematic of Indirect-TCP

Fast Retransmission

This approach was suggested by Caceres and Iftode [8] to overcome the delay in transmissions caused by intermittent disconnections such as those that occur when a mobile host (MH) moves to a foreign agent (FA) during a TCP communication. As we have already discussed, TCP transmission behavior after a disruption depends on the duration of the disruption. Extremely short disruptions (lasting for a time much less than RTO) appear as a short burst of packet losses. In this case, the TCP retransmits those packets for which time out occurs and recovers without slow start. However, for long disruptions (lasting for a time much greater than RTO), TCP resorts to slow start. This results in inefficiency. As soon as the mobile host registers at a foreign agent, it starts sending duplicate acknowledgments. As is standard with TCP, three consecutive acknowledgments for the same TCP segment are inferred as a packet loss by the host-end TCP, but it is also inferred that the connection is live, thereby causing the fast retransmit behavior of TCP.

The advantage of this scheme is that it reduces the disruption time after the MH is reconnected, otherwise the FH would wait for RTO unnecessarily. The disadvantage of this approach is that it does not propose a general approach for TCP communication in mobile wireless networks. For example, it does not address the specific error characteristics of the wireless medium.

Snooping TCP

Balakrishnan et al. [4] proposed a protocol that improves TCP performance by modifying the software at the base station while preserving the end-to-end TCP semantics. The modified software at the base station is known as "snoop." It monitors every packet that passes through the TCP connection in both directions, that is, from the MH to the FH and vice versa. It buffers the TCP segments close to the MH. When congestion is detected during the sending of packets from the FH to MH in the form of duplicate acknowledgments or time out, it locally retransmits the packets to the MH if it has buffered the packet and hides the duplicate acknowledgments.

An advantage of snooping TCP is that it maintains the TCP semantics by hiding the duplicate acknowledgment for lost TCP segments and resends the packets locally. However, it also suffers from higher overhead incurred when the MH moves from current the BS to new base station; the packet buffered at the current base station need not be transferred to the new base station.

Mobile TCP

This protocol was proposed by Brown and Singh [7]. In mobile wireless networks, users would suffer badly from unacceptable delays in TCP communications and frequent disconnections caused by events such as signal fades, lack of bandwidth, handoff, unless these are explicitly handled by the protocol. Mobile TCP (M-TCP) tries to avoid the sender window from shrinking or reverting to slow start when bit errors cause packet loss, as was attempted in I-TCP and snooping TCP. In this protocol, as in I-TCP, the TCP connection from fixed host to the mobile host the segmented into wired and wireless parts. The wired connection is from fixed host (FH) to the supervisory Host (SH) and the wireless connection is from SH to the mobile host (MH). Many MHs are connected to the SH through several base stations, as shown in Figure 2.8. The SH supervises all the packets transmitted to the MH and acknowledgments sent by the MH. It is also used as an interface from the FH

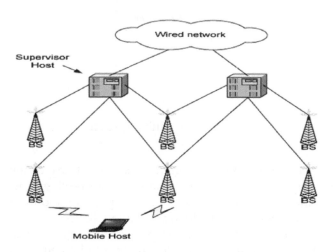

Figure 2.8: Schematic of the operation of M-TCP

to the MH and vice versa.

When the packet sent to the FH by the MH using the SH, the wired part uses the normal unmodified TCP and the wireless part uses a modified version of TCP known as M-TCP to deliver data to the MH. This packet is acknowledged only when the MH receives the packet. Thus it maintains the TCP semantics, unlike I-TCP. If the acknowledgment is not received by the FH, the SH decides that the MH is disconnected and sets the sender FH window size to zero. This prevents retransmission. When the SH notices that the MH is connected, it sets the full window size of sender FH. When the MH moves from the current SH region to the new SH region, a state transfer takes place, so that the new SH can maintain a TCP connection from the FH to the MH.

2.9   Networking Devices

Networking devices are required to connect some hosts together to form a network and to interconnect different networks together to make an inter-network. Various kinds of devices are used. Here we discuss some of the basic devices important for network communication.

Hub

A hub is a device that has many RJ-45 ports to which many computers can be connected, as shown in Figure 2.9. Typically, computers are connected to a hub using twisted-pair cabling. A hub relays any signal that it receives at any of its ports to all other ports. Thus, in essence, it serves as a broadcasting device. A hub is said to operate in the physical layer. It is called a physical

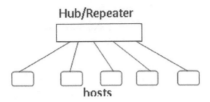

Figure 2.9: Hub/repeater

layer device, as it transmits the signal it receives at any of the ports to all other remaining ports and does not check any addressing or formatting aspects of the data stream it receives. A hub is a popular and inexpensive way to connect computers and embedded computing devices. In addition to ports for connecting computers, even an inexpensive hub generally has a port designated as an uplink port that enables the hub to be connected to another hub to create larger networks. As can be seen from Figure 2.9, the local area network obtained using a hub has a star topology. The hub is at the center of the star and broadcasts any signal it receives from any sending node.

Advantages:

- It regenerates the signal before sending it to others, which helps to avoid signal attenuation problems. In other words, it effectively increases the maximum distance that is achievable between nodes.

- It is inexpensive as compared to other network devices.

- As it does not perform any processing of the signal other than broadcasting, it operates faster than other networking devices.

Limitations:

- It is not a scalable means to realize a LAN. That is, one cannot use a large number of hierarchically connected hubs to obtain a large LAN. The reason is that, since the hubs are broadcasting devices, as the number of nodes increases in the LAN, the number of collisions increases drastically. In other words, hubs do not help to restrict the collision domain. A consequence of this is that the available bandwidth is shared among the hosts in the LAN. That is, when a large number of nodes are connected using hubs to form a LAN, the effective bandwidth available to each node is substantially reduced, and the response time increases.

Switches

A switch connects multiple hosts. A switch operates in both the physical and the data-link layers and is more intelligent than a hub. It regenerates the signal it receives. Additionally, it stores the data before forwarding and constructs the frame from the data stream to determine the source and destination MAC addresses. However, switches are at present available that do not have to store a frame to determine its MAC address and are called cut-through switches.

A switch is much more intelligent than a hub, since, unlike a hub which broadcasts the signal it receives, a switch forwards the frame only along the link that connects to the host to which it is addressed. As a result, the total network traffic in the LAN decreases, and it effectively reduces the number of collisions as compared to a hub. Further, switches can provide additional improvement of performance over a hub as the transmissions are full duplex. That is, a node can send and receive data at the same time. Full duplex transmission can be obtained, as the switch effectively isolates the sending and receiving signals for a node.

Advantages:

- Avoids signal attenuation problems by regenerating the signal.

- Processes each frame and sends only to the destination host. Therefore it in effect reduces the collision domain in a LAN.

- Simultaneous connections through the switch are possible. That is, at the same time multiple pairs of hosts can communicate along different pairs of links. This, in effect, increases the effective bandwidth as compared to a hub.

Limitations:

- Typically takes more processing time than a hub, as it needs to interpret the MAC address.

- It is more costly as compared to a hub.

Routers

A router is used to interconnect more than one network. As shown in Figure 2.10, network1 and network2 are connected using a router. Each network may be formed by a set of hosts connected to a hub/switch. It is a three-layer device; it operates in the physical, data-link, and network layers. A router operates in the network layer, as it checks the destination IP address to determine the link along which to forward the packet. It dynamically builds a forwarding table using routing algorithms like link-state or distance-vector. When it receives a packet through a port, it inserts it into a queue. It processes packets from the front of the queue. As a part of processing, it checks the IP address of the destination and finds the next hop for that destination from the forwarding table, after which the router executes ARP to get the next hop's MAC address and forwards the packet to the next hop.

Wireless Access Point

A Wireless Access Point (WAP) is a transmitter and receiver of radio signals. Therefore a WAP has a built-in antenna, transmitter, and receiver circuitry.

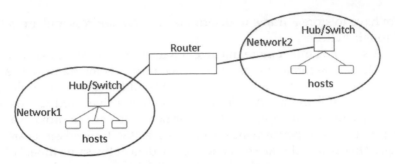

Figure 2.10: TCP/IP stack

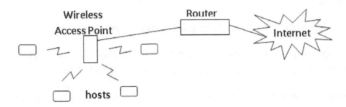

Figure 2.11: Wireless connection to the Internet

A WAP is typically used to provide a computer with a wireless networking capability to connect to a wireless LAN or a wired LAN. A WAP allows several hosts to communicate with the network via this device. It can also allow two hosts to communicate with each other through it. In this case, it works as a switch. A WAP provides flexibility to users in the sense that it allows them to move within the wireless range of the access point. As shown in Figure 2.11, an access point receives packets from hosts and forwards them to the Internet via the router; when received from the router, it broadcasts the packet into the air so that the intended receiver gets the packet.

Summary

In this chapter, we first reviewed some basic concepts regarding computer communication protocols. We briefly discussed the OSI protocol, which serves, a pedagogical purpose and the TCP/IP protocol suite, which has been accepted as the de facto protocol for the Internet and a host of other applications. We then identified the important problems that might arise when TCP is used as it is in mobile wireless networks. We discussed a few important adaptations that have been proposed to make TCP work satisfactorily in the mobile environment. Finally, some basic networking devices along with their operations were discussed.

Exercises

1) State whether the following are True or False. Give a brief reason for your answer in each case.

    a. TCP is a peer-to-peer, connection-oriented protocol.

    b. A link layer is responsible for delivering data packets by making use of lower layer protocols and also provides error detection and packet framing functionalities.

    c. TCP guarantees that data will be delivered without loss, duplication, or transmission errors.

    d. TELNET encodes multimedia data using the MIME protocol.

2) Explain the following terms associated with the TCP/IP stack:

    a. IP

    b. HTTP

    c. SMTP

    d. MIME

    e. FTP

    f. SNMP

    g. ICMP

    h. ARP

    i. RARP

    j. DNS

    k. IP Addresses

    l. IGMP

3) Write short notes on

    a. TELNET

    b. FTP

    c. SMTP

    d. TCP/IP versus the ISO/OSI protocol model

4) Explain the layered architecture of the TCP/IP protocol suite and compare it with the ISO/OSI architecture.

5) Answer the following with respect to missing and duplicate segments in TCP operation.

    a. What can cause segments to be missed at the receiver end and also cause duplicate segments to arise? Explain your answer using a suitable scenario of operation.

    b. How exactly is a missing segment detected in TCP? Explain the specific actions that take place when a missing segment is detected.

6) What is slow start in TCP operation? Explain how it works. How does slow start help improve the performance of TCP?

7) What problems would occur if the traditional TCP is used in mobile wireless environments? Discuss how TCP can be adapted to work efficiently in a mobile network environment.

8) Explain indirect TCP with the help of a suitable schematic diagram.

9) Why are the I-TCP acknowledgments and semantics not end to end? What are the implications of this?

10) What is the snooping TCP approach in mobile wireless networks? Discuss its advantages.

11) How are handoffs handled in snooping TCP?

12) Briefly discuss the M-TCP approach of extending TCP to work efficiently in mobile wireless networks. How does M-TCP maintain end-to-end semantics?

13) Why does congestion occur in a network? Explain how TCP detects and handles congestion.

14) It is said that in TCP packet losses occur primarily due to congestion. Explain why packet losses due to reasons such as packet corruption are rare even when the medium is noisy.

15) Compare hub and switch. Which one is preferred and why?

16) What is the function of a router?

Bibliography

[1] A. Ahuja, S. Agarwal, J. P. Singh and R. Shorey, "Performance of TCP over Different Routing Protocols in Mobile Ad-hoc Networks," IEEE Vehicular Technology Conference 2000, vol. 3, pp. 2315–2319, Tokyo.

[2] Xiang Chen, Hongqiang Zhai, Jianfeng Wang and Yuguang Fang, "A Survey on Improving TCP Performance over Wireless Network," Dept. of Electrical and Computer Engineering, University of Florida, Gainesville, FL.
http://winet.ece.ufl.edu/ zhq/book05chen1.pdf.

[3] V. Anantharaman, S.-J. Park, K. Sundaresan and R. Sivakumar, "TCP Performance over Mobile Ad-hoc Networks: A Quantitative Study," Wireless Communications and Mobile Computing Journal (WCMC), Special Issue on Performance Evaluation of Wireless Networks, 2003.

[4] H. Balakrishnan, V. Padmanabhan, S. Seshan and R. Katz, "A Comparison of Mechanisms for Improving TCP Performance over Wireless Links," *Proceedings of ACM SIGCOMM'96*, Aug. 1996.

[5] A. Bakre and B. R. Badrinath, "I-TCP: Indirect TCP for Mobile Hosts," *Proc. 15th International Conf. on Distributed Computing Systems (ICDCS)*, May 1995.

[6] P. Bhagwat, P. Bhattacharya, A. Krishna and S. K. Tripathi, "Enhancing Throughput over Wireless LANs Using Channel State Dependent Packet Scheduling," *IEEE INFOCOM'96*, San Francisco, March 1996.

[7] K. Brown and S. Singh, "M-TCP: TCP for Mobile Cellular Networks," *ACM Computer Communication Review*, vol. 27, no. 5, Oct. 1997.

[8] R. Caceres and L. Iftode, "Improving the Performance of Reliable Transport Protocols in Mobile Computing Environments," *IEEE JSAC.*, vol. 19, no. 7, July 2001.

[9] K. Chandran, S. Raghunathan, S. Venkatesan and R. Prakash, "A Feedback-Based Scheme for Improving TCP Performance in Ad Hoc Wireless Networks," *IEEE Personal Communications*, 8 (1):34-39, February 2001.

[10] T. D. Dyer and R. V. Boppana, "A Comparison of TCP Performance over Three Routing Protocols for Mobile Ad Hoc Networks," *ACM Mobihoc*, October 2001.

[11] Z. Fu, P. Zerfos, H. Luo, S. Lu, L. Zhang and M. Gerla, "The Impact of Multihop Wireless Channel on TCP Throughput and Loss," *IEEE INFOCOM'03*, San Francisco, March 2003.

[12] G. Holland and N. H. Vaidya, "Analysis of TCP Performance over Mobile Ad Hoc Networks," *MOBICOM'99*, Seattle, August 1999.

[13] Behrouz A. Forouzan, *"Data Communication and Networking,"* 4th Edition, McGraw-Hill, 2007.

CHAPTER 3

NETWORK PROGRAMMING USING SOCKET API

3.1 Introduction

Network programming refers to designing programs that may run on different machines but are capable of communicating with each other. That is, the programs cooperatively solve some problem by communicating with each other. Unless otherwise specified, the words *program, process,* and *application* are used in this text synonymously, and the term communication refers to unicast communication, in which communication takes place between one sender and one receiver. Unicast communication is also called one-one communication. We consider a message to be the basic unit of communication. The different possibilities as to where these communicating programs/processes can reside are as follows:

i) All reside in a standalone computer.

ii) Each resides in a different computer and the computers are connected through a network.

iii) Each resides in computers that are in different networks connected to Internet.

The elegance of the TCP/IP stack is that programs, once developed, can communicate and work satisfactorily regardless of on which computers the program components are hosted. The only requirement is that the computers hosting the program components must be connected via some network. The programs may even work satisfactorily when the program components are hosted on computers spread across the globe when connected via the Internet. This type of program is also called *Client-Server* computing technology. The client-server programming paradigm is schematically shown in Figure

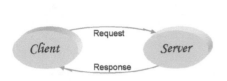

	Server	Client
	Application Layer	Application Layer
	Socket	*Socket*
	Transport Layer	Transport Layer
	Network Layer	Network Layer
	Datalink Layer	Datalink Layer
	Physical Layer	Physical Layer

Figure 3.1: Client-server paradigm

Figure 3.2: A socket as an interface
and end points of a logical channel

3.1, in which a client makes a request and the server responds to it. Several examples of client-server programs have been shown in Table 3.1. Note that the client and server refer to two different programs. Network programs are convenient to model in terms of pairs of programs, in which one program that initiates the communication process corresponds to the client and the one that responds to the client corresponds to the server. However, in a general client-server program, many clients may communicate with one or more servers.

The objective of this chapter is to help to develop network applications at the top of the transport layer through the socket interface as shown in Figure 3.2. There are many standard applications available for day-to-day network services. Some of these are quite popular and are used every day. Examples of a few popular applications are given in Table 3.1. These are all precompiled programs and may be used to communicate across a network. In this chapter we focus our discussions on developing client-server applications. Although this kind of program can be developed in any platform and by using a variety of programming languages, the discussion in this book is restricted to C programming on the Linux platform only. The socket API (Application Programming Interface) is used in C programs to develop network applications.

Table 3.1: Popular network applications

Client Program	Server Program
Web client	Web server
(Example: iexplorer, mozzila)	(Example: IIS, Apache, Tomcat)
FTP Client	FTP server
Telnet Client	Telnet server
E-Mail client	E-Mail server

3.2   Socket Interface

Socket is a software abstraction that is used as an interface between application and transport layers of the TCP/IP stack (Figure 3.2). Just as an electrical

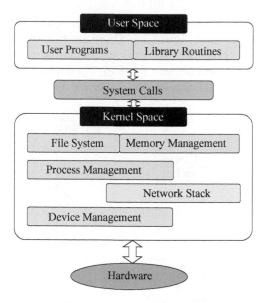

Figure 3.3: Linux kernel vs. user space

socket is used to send/ receive electrical power, the network socket is used to send/ receive information. Sockets are created in pairs, i.e., one on client side and another on the server side. Two sockets work as two end points of a logical communication channel between the sender and the receiver. This channel is analogous to a pipe used for water flow.

In Unix systems the socket is implemented as a file. Like any ordinary file, it also returns a file descriptor when created. This descriptor is usually termed the socket descriptor. When the server has some data to send to the client, it writes the data to its own socket interface and the information becomes automatically available at the client-side socket, from which the client can read it. However, it requires that the underlying network is in place and the transport and other layers below it perform their responsibilities satisfactorily on both the sides. Socket API provides a library with a rich set of system calls, utility functions, and data structures to develop network applications. There are many kinds of sockets available for many different purposes, and this chapter discusses only Internet sockets. For other kinds of sockets, the reader is referred to the standard literature on socket programming, some of which are listed in the bibliography section. The complete network stack of protocols is implemented in the native operating system kernel along with other operating system features. The kernel usually runs in a restricted area of the memory called the kernel space, and user programs are executed in another part of the memory called the user space. Direct access to kernel by the user is restricted to avoid corruption of the kernel data structure that might cause a crash to the system. Therefore system calls are used as an interface between the kernel space and the user space. A simplified view of a Linux kernel is shown in Figure 3.3. There are two types of Internet sockets, the stream socket (provides connection-oriented service) and the datagram socket (provides connection-less service). The former uses TCP and the latter uses UDP as the underlying protocol. Figure 3.4 provides a block diagram

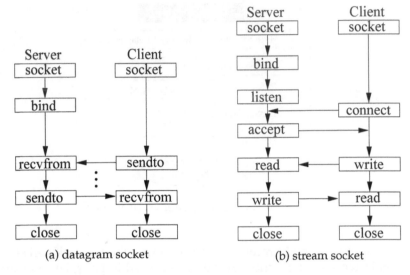

(a) datagram socket   (b) stream socket

Figure 3.4: Block diagram of socket program structure

for a basic network program. For a datagram socket (UDP), the system calls are executed in the sequence shown in Figure 3.4(a) and stream sockets (TCP) use the sequence in Figure 3.4(b). All socket programs essentially incorporate the steps depicted in either of these block diagrams.

The sequence of execution for the ideal case (no error) of a stream socket program is given in Table 3.2, and the corresponding C program code for the server is provided in Listing 3.1. The server program in Listing 3.1 receives only one message from the client Then it stops after printing the received message. We explain this program in some detail in the next section. However, let us first compile and execute the program. The programs given in this text are executed for Fedora and Ubuntu distributions of Linux.[1] Follow the steps to compile and execute the network program under any Linux distribution.

Step 1.   Use any text editor to write the codes given in code Listing 3.1 and save as server.c.

Step 2.   Open a command terminal and issue the following command to compile the server program:

---

[1]Some minor modifications may be required for other platforms.

Table 3.2: Client-server execution sequence

Server Program	Client Program
1. Create a server-side socket	1. Create a client-side socket
2. Bind the server socket to a local socket address (port, ip)	
3. Listen for an incoming connection request	
	4. Connect request made by the client
5. Connect request accepted by the server	
6. Talk to the client	6. Talk to the server
	7. Close the client socket when the conversation is finished
8. Optionally close the server socket	

        ```
$] cc -o server server.c
```
If there is no syntax error, then it produces an executable file with the name *server*. Otherwise clear the syntax error according to the compilation report repeatedly until all the errors are cleared.

Step 3   If, in the previous step, the program compiles successfully, then issue the following command to run the server program:

        ```
$] ./server
```
If executed successfully, it will print the the server port number (a positive integer) in the terminal, as shown in Figure 3.5. Record this server port number and also note the server IP address.

If both the terminals are running in the same computer then the IP address of client and server will be the same.

Step 4.   Open another command terminal in either the same or a different computer in the same LAN. The former terminal acts as the server and the new terminal acts as the client.

Step 5.   In the client terminal, issue the following command:

        syntax: telnet <server-IP>   <server-port>

    ```
$] telnet 127.0.0.1   56808
```

If the telnet client has executed successfully, then it prints the first three lines in the client terminal, as given in Figure 3.6. Now enter the string "First message to Server!!!" from the keyboard in the telnet client terminal (Figure 3.6), in the next moment the message is printed in the server terminal (Figure 3.5). After printing the message from the telnet client, the server stops (line 37–40) and therefore the client also stops. Now run the program a few times with different server IP and port numbers and with different messages.

Listing 3.1: TCP Single Message Server

```
1  /* This is a server program: server.c
2  * compile with : cc -o server server.c
3  * execute with : ./server */
4  #include <netinet/in.h>
5  #include <stdio.h>
6  #include <string.h>
7  #include <stdlib.h>
8
9  int main() {
10   int sockfd,connfd;
11     char buf[1024];
12     struct sockaddr_in server;
13     struct sockaddr_in client;
14   socklen_t cliAddrLen=sizeof(client), servAddLen;
15
16   /* (1)Create Socket */
17   sockfd =socket(AF_INET,SOCK_STREAM, 0);
18   if(sockfd < 0) {
19         perror("\nError in opening socket ... ");
20         exit(1);
21   }
22
23       server.sin_family = AF_INET;
24   server.sin_addr.s_addr = htons(INADDR_ANY);
25   server.sin_port = 0;
26
27   /* (2)Bind server socket to a local protocol address*/
28     if(bind(sockfd,(struct sockaddr *)&server, sizeof(server))){
29             perror("\nError in bind ... ");
30             exit(2);
31     }
32
33   servAddLen=sizeof(server);
34     if(getsockname(sockfd, (struct sockaddr *)&server, &
     servAddLen)) {
35                 perror("\nError in getting port... ");
36                 exit(3);
37     }
38
39   printf("\nSocket has port # %d \n", ntohs(server. sin_port));
40
41   /* (3)Listen for a incoming connection request.*/
42     listen(sockfd, 5);
43
44   /*(5)Accept the request from client*/
45     connfd = accept(sockfd, (struct sockaddr *)&client, &
     cliAddrLen);
46     if(connfd == -1){
47           perror("\nError in accept... ");
48           exit(0);
49     }
```

```
50
51    /* (6) Talk to Client */
52      memset(buf, 0, 1024);
53      recv(connfd, buf, 1024, 0);
54      printf("\nMessage from client :: \n \%s\n", buf);
55
56    /* (7) Close Sockets */
57      close(connfd);
58      close(sockfd);
59  }
```

Screen shots of the server terminal are provided in Figure 3.5 and for the client in Figure 3.6. The IP address used in the above execution (127.0.0.1) is called a local loop address. When you want to communicate between the client and the server programs residing in the same computer, actually you don't even need a network interface because in this case the message never goes outside the computer. Therefore the above IP address is reserved for this kind of communication, where a computer needs to communicate with itself and the messages loop back to the same system. However, if you are running the telnet command from a different machine, then, while running the client program, provide the correct IP address of the server machine instead of 127.0.0.1.

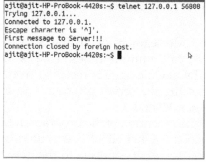

Figure 3.5: Running the server program

Figure 3.6: Running the client (telnet) program

Let us try to understand Listing 3.1 with respect to the server to side block diagram given in Figure 3.4. Lines 4−7 and directives to include libraries that contain the data structures and system calls required for the program. An explanation of various functions and data structures is given in the following section. Lines 10 to 14 declare certain required variables such as descriptors, message buffer, and the socket address structure. Lines 17 to 25 create a socket and initialize the values for the socket address structure. In lines 28−39, the created socket is bound with the IP/port number and the port number to be used by the client is printed out. Line 42 makes the server ready to listen to any incoming client request for connection. Lines 45 to 49 complete the connection request by acknowledging the client. If everything goes fine, then the connection between the client and the server is established and the server is ready to receive messages from the client. Lines 52 to 54, pertain to receiving a message from the client and displaying it. Finally, lines 57 to 58 close the connection. Line 59 is the end of the program.

Now refer to lines 53 to 58, of Listing 3.1. The uses one recv() call followed by close() calls. Therefore the server closes after receiving one message from the client, causing the client to close and get disconnected.

Let us now modify lines 53 to 54 of Listing 3.1 with the lines of codes provided in Listing 3.2 to receive more than one message from the client.

Listing 3.2: Multimessage TCP Server

```
53 do {
54   len = recv(connfd, buf, 1024, 0);
55       if (len > 0){
56     printf("\n Message from client :: \n%s\n", buf);
57     memset(buf, 0, 1024);
58   }
59   else
60     printf("\nClient Exited...\n");
61 }while(len > 0);
62
63 /* Close Sockets */
64 . . .
```

The code segment of Listing 3.2 causes the server program to execute the recv() call multiple times. That is, the do-while loop continues to execute until some message is received from the client. Now follow the same steps to modify the server program and execute it to obtain the result shown in Figures 3.7 and 3.8

The execution of the program in Listing 3.2 is given in Figures 3.8 and 3.9. When the telnet client is executed by providing the IP and port addresses of the server, it connects to the new server. Now messages can be sent one by one and the server goes on printing all of these messages in sequence. In this run (Figures 3.7 and 3.8), client sends three messages and the server prints all of them. After the third message, telnet is disconnected by entering the escape sequence (ctrl+]) followed by the quit command. The server also stops by displaying "Client Exited." Now look at the modified code; recv() call returns the size of the message received or 0 on error. When the telnet client quits, recv() returns 0 and the do-while loop terminates and the server stops.

We have used the precompiled telnet client program usually available with all operating system installations. However, it is also possible to design our own client program so that, in place of the predesigned telnet client, we can use our customized client. Listing 3.3 gives a simple client program that sends one message to the server and quits.

Figure 3.7: Server terminal receiving multiple messages

Figure 3.8: Telnet client sending multiple messages

Listing 3.3: TCP Client program: client.c

```
1  #include<sys/types.h>
2  #include<sys/socket.h>
3  #include<netinet/in.h>
4  #include<stdio.h>
5  #include<stdlib.h>
6
7  main (int argc ,char *argv[ ]){
8     int soc,addrlen;
9     struct sockaddr_in  server;
10    char buf[1024];
11
12       /* Create Socket */
13    soc = socket(AF_INET, SOCK_STREAM, 0);
14    if (soc < 0){
15        perror("\nError in openig socket");
16        exit(1);
17    }
18
19       /* Assign server protocol address */
20    server.sin_family = AF_INET;
21    server.sin_addr.s_addr = inet_addr(argv[1]);
22    server.sin_port = htons(atoi(argv[2]));
23
24       /*Connect request to server */
25    if(connect(soc, (struct sockaddr*)&server, sizeof(server)) <
       0){
26      perror("\nError in connection");
27      exit(2);
28    }
29
30       /* Talk to Server */
31    printf ("Enter a messgage for server");
32    scanf("%s",  buf);
```

```
33    write(soc, buf, sizeof(buf));
34
35    /* Close the socket*/
36    close(soc);
37  }
```

Now follow the steps below to run the client program corresponding to the server programs given in Listings 3.1 and 3.2.

Step 1. Write the client code as given in Listing 3.3 and save it as client.c.

Step 2. To compile the client, issue the following command:

```
$ cc -o client client.c
```

Step 3. Execute the server as before and record the port number.

Step 4. At the client terminal issue the following command:

```
$ ./client <server-IP>  <server-port>
```

If the client executes successfully, then it asks to enter a message for the server. Enter any string and as soon as you press the enter key, the message is printed in the server terminal and both client and the server programs terminate.

Make some investigations by running the server and the client programs from terminals opened at different computers of a LAN. That is, you can ask your friend to run the server program on some other machine and get its port number. Now run the client from your computer using a friend's port number and then talk with each other using text message chatting.

Test/Viva Questions

a) What is a socket interface? Give an example application in which it is used.

b) What is unicast communication? What are other types of communications?

c) Name the layers in the TCP/IP stack from the bottom to the top.

d) What do you understand by client server programming? Give examples of some client and server applications.

e) Why are network programs written as pairs of programs?

f) Socket interfaces work in which layer of the TCP/IP stack?

g) What is an IP address? How many bits specify an IPv4 address?

h) Distinguish between a stream socket and a datagram socket. What are their practical uses?

i) Draw a block diagram to show the sequence of system call usage in a client/server program that uses a stream socket.

j) Draw a block diagram to show the sequence of system call usage in a client/server program that uses a datagram socket.

Programming Assignments

1) Modify the server code in Listing 3.1 such that it never stops (runs forever) and it can receive multiple messages from a telnet client. When one client is disconnected, another client can connect to the same server and communicate. If you want to stop the server, then you would have to stop it manually (ctrl+c).

2) Modify the client code in Listing 3.3 such that it sends multiple messages to the server. The client should stop when the user enters the message "bye." Modify the server program in Listing 3.2.

3) Write client/server programs such that the server echoes the same messages back to the client that it receives from the client (echo server).

4) Write a client/server program such that the server sends its system time to the client. (time server)

5) Write client/server programs such that both can send to and receive from each other.

3.3   Socket API

So far we have concentrated on the overall structure of client server programs. Now it is time to understand each line of the programs. For this reason we need to understand the socket API. As the reader might have noticed, there are some words used in the program code which are normally not found in ordinary C programs. Before going further, we first provide the full form of these terms in Table 3.3.

3.3.1   Data structures

In this section, we first explain the data structures used to represent IPv4 addresses. Subsequently, we discuss the IPv4 socket address structure.

1. **IPv4 Address Structure**

An IPv4 address is a 32-bit integer that is used to locate a host in a network. Any computer connected to a TCP/IP network needs at least one unicast IP address. To store this address, the following structure is used.

```
#include<netinet/in.h>
struct in_addr{
     in_addr_t s_addr;
     };
```

Table 3.3: Some terminology related to socket API

Terminology	How to read
AF	Address Family
PF	Protocol Family
INET, in	Internet
SOCK, s	Socket
sockaddr, sa	Socket Address
sin	Internet Socket
INADDR	Internet Address
uint	unsigned integer
t	type
hton	host to network
ntoh	network to host

where some types are defined as
//32 bit unsigned net/host-id in network byte order

```
typedef uint32_t in_addr_t;
```
//16 bit unsigned port-id in network byte order

```
typedef uint16_t in_port_t;
```
similarly one constant is declared as
//a wildcard (any) IP address

```
#define INADDR_ANY((in_addr_t) 0x00000000)
```
The use of this wildcard address is discussed later in this section. As the IP address is an integer, it may be represented in different ways, known as byte ordering.

Byte Orderings

The API supports two different byte orders to represent an IP address.

i) *Network byte order* is the order in which a multibyte integer is represented in a network. It is an integer in big-endian format, i.e., most significant byte (MSB) first form.

ii) *Host byte order* is the order in which a multibyte integer is represented in a device. It may be in either little-endian or in big-endian format. For example, Intel's Core i7 processor uses little-endian, whereas Motorola's latest ColdFire(R) processor uses big-endian formats.

If the components of an IP address 179.100.35.6 are stored in the same order, this is called big-endian (MSB-first). But if it is stored in reverse

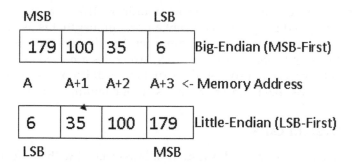

Figure 3.9: Byte orderings

order as 6.35.100.179, then it is called little-endian least significant byte (LSB) first. The different byte orderings are given in Figure 3.9.

Different hosts may use different byte orderings. When these hosts communicate with one another over the network, then, due to different byte orderings, the same IP address or port number may be interpreted differently in the two systems, thereby making any communication impossible. The programing interpreting the byte stream needs to be switched for different byte orders, which is a formidable task. This is because it may not be possible to know the host byte order beforehand. Therefore, to improve interoperability, API provides some utility functions to convert between host and network byte orders. Application developers simply need to convert the IP/port numbers to the network byte order to store in memory and to convert to host byte order to print the addresses (if required). The different byte ordering functions are provided in a separate subsection.

2. **IPv4 Socket Address Structure**

   Information regarding a socket is stored in another structure called the socket address structure. The different fields of this structure are given as follows.

```
#include<netinet/in.h>
struct sockaddr_in{
    uint8_t sin_len;
    sa_family_t sin_family;
    in_port_t sin_port;
    struct in_addr sin_addr;
    char sin_zero[8];
    };
```

Where *sin_len* represents the length of the structure. The length of the structure can be computed to be 1 + 1 + 2 + 4 + 8 = 16 bytes. *sa_family_t* is the address family of the socket address structure, *in_port_t* is a 16-bit unsigned integer to represent a port number that is used to identify an application running on a computer. The *sin_zero* member is unused, but it is set to zero. Socket address structures are used only on a given host to represent a socket; the structure itself is not communicated but some fields like IP and port are used for communication.

3.3.2   System calls

In this section we describe the system calls associated with socket communication processes.

a) **socket()**

```
#include<sys/socket.h>
int socket( int family, int type, int protocol);
```
The *socket()* call is used to create sockets in both client and server programs. On success, the socket function returns a nonnegative integer, known as the *socket file descriptor*. The argument *family* specifies the protocol family. There are different families available, but in this text, only *AF_INET* (Internet address family) is used. The argument *type* defines the communication type; for connection oriented (TCP) *SOCK_STREAM* is used, and *SOCK_DGRAM* is used for connection less datagram sockets (UDP). Another type, *SOCK_RAW*, is used for raw sockets. The *protocol* argument is used to represent the lower transport layer protocol that will use the socket service. The argument may be set to any one of the transport layer protocols from Table 3.4 or is set to 0, so that socket chooses the default protocol based on socket type. That is, TCP is chosen for a stream socket type and UDP for a datagram socket type. Lines 17 to 21 of Listing 3.1 and lines 13 to 17 of Listing 3.3 create the corresponding sockets. The sockfd and soc variables are used to store the socket descriptors in the server and the client programs, respectively.

Table 3.4: Protocol types

Protocol	Description
IPPROTO_TCP	TCP transport protocol
IPPROTO_UDP	UDP transport protocol
IPPROTO_SCTP	SCTP transport protocol

b) **bind()**

```
#include<sys/socket.h>
int bind(int sockfd, const struct sockaddr *myaddr,
```

```
socklen_t addrlen);
```
The *bind*() call assigns a local protocol address to a socket. A local protocol address consists of an IP address along with a port number. This call is used in the server and is optional in the client. The first argument is the socket descriptor, the second argument is a pointer to the address structure, and the third argument is the length (size) of the address structure (32-bit integer). It returns 0 if OK and -1 on error. Normally, a TCP client does not bind an IP address to its socket. The kernel chooses the source IP when the socket is connected, based on the outgoing interface.

When the *bind*() call is used for a TCP socket, then the call lets us specify either a port number or an IP address, both, or neither. If the user does not specify a value for either both or none of them before the bind call, then the operating system kernel chooses the correct values inside the bind call. To let the kernel decide these values, we need to write lines 23 to 25 of Listing 3.1 before the bind call. Note that, if the port number is assigned zero and/or IP address IN_ADDR_ANY, then the operating system kernel chooses some valid values for the port number and the IP address. But if some values are assigned to these variables before the bind call, then the kernel uses the same values in the bind call. Since the server programs running on a computer are distinguished by their port number, care should be taken not to assign the same port numbers to different server programs running on the same computer.

To get the values of PORT and IP addresses from the kernel space to the user space, the function '*getsockname*() is used, as given in Line 34 of Listing 3.1. This function is discussed in Section 3.3.5.

c) **listen()**

```
#include<sys/socket.h>
int listen(int sockfd, int backlog);
```
The *listen* function converts an unconnected socket into a passive socket, indicating that the kernel should accept incoming connection requests directed to this socket. The call to listen moves the socket from the CLOSED state to the LISTEN state. The second argument specifies the maximum number of connections that the kernel should queue for this socket, as given in line 42 of Listing 3.1. This call is used in the server programs only.

d) **connect()**

```
#include<sys/socket.h>
```
*int connect(int sockfd, const struct sockaddr *servaddr, socklen_t addrlen);*
This call is used by a client to send a connection request to a server. The *connect*() call initiates TCP's three-way handshaking process for

connection establishment with a TCP server. The function returns only when the connection is established (0) or when an error occurs (-1). If the connection request fails, the socket is no longer usable and must be closed.

e) **accept()**

```
#include<sys/socket.h>
int accept(int sockfd, struct sockaddr *cliaddr,
socklen_t *addrlen);
```
This is called by a TCP server to return the next completed connection from the front of the connection queue. If the completed connection queue is empty, the process is put to sleep. The *cliaddr* and *addrlen* arguments are used to return the protocol address of the connected peer process. The *addrlen* contains the size of the client address structure, and on return it contains the actual number of bytes stored by the kernel in the socket address structure.

If accept is successful, it returns a new descriptor that was automatically created by the kernel. This new descriptor refers to the TCP connection with the client. Now the first descriptor created by the socket call (discussed earlier) is known as the *listening socket*, and the second descriptor created by the accept call is known as the *connected socket*. A given server normally creates only one listening socket, which exists for the lifetime of the server. The kernel creates one connected socket for each client connection that is accepted. When the server finishes serving a client, the connected socket for that client is closed. Line 45 of Listing 3.1 demonstrates the use of the accept call.

This function returns three values:

i) An integer returned directly. This value represents either a new socket descriptor (if +ve) or an error indication (-1).

ii) The protocol address of the client process, returned indirectly through the reference argument *cliaddr*.

iii) The size of this address is returned indirectly through the reference argument *addrlen*.

If the protocol address is not required, then both *cliaddr* and *addrlen* are set to NULL;

f) **close()**

```
#include<unistd.h>
int close(int sockfd);
```
The default action of this call is to mark the socket as closed. After the call to close() is made, the socket descriptor is obviously no longer usable by the process. However, any data that is already queued would

be sent to the other end, and after this the normal TCP connection termination sequence takes place.

f) **Message Sending and Receiving**

Different functions are available for stream and datagram services as given below.

Receiving Calls

`int read(int sockfd, char *buf, int nbytes);`
This call helps read *nbytes* (max) of data from the socket to the buffer. It returns the number of bytes read successfully from the socket and 1 on error. This call is normally used for TCP.

`int recv(int sockfd, char *buf, int nbytes, int tag);`
This reads *nbytes* (max) from the socket to the buffer. It returns the number of bytes read successfully from the socket and 1 on error. An extra tag is used match the *send()* with *recv()*, and this call is used in TCP.

`int recvfrom(int sockfd, char *buf, int nbytes, int flag, struct sockaddr *from, int addrlen);`
This receives *nbytes* (max) from a socket, whose address is given by the from address structure to the buffer. It returns the number of bytes read successfully from the socket and a 1 on error. It is normally used in UDP.

Sending Calls

`int write(int sockfd, char *buf, int nbytes);`
This writes *n* bytes (max) to another socket from the buffer. It returns the number of bytes written successfully to the socket and 1 on error. It is normally used with TCP.

`int send(int sockfd, char *buf, int nbytes, int tag);`
This sends *n* bytes (max) to the other connected socket from the buffer. It returns the number of bytes written successfully to the socket and 1 on error. It is normally used with TCP.

`int sendto(int sockfd, char *buf, int nbytes, int flag, struct sockaddr *to, int addrlen);`
This sends *nbytes* (max) to the other socket, whose address is given by the address structure from the buffer. It returns the number of bytes sent successfully from the socket and a −1 on error. It is normally used with UDP.

All receiving calls behave the same if they are called as follows:

```
        read(sockfd, buf, len);
or
        recv(sockfd, buf, len, 0);
or
        recvfrom(sockfd, buf, len, 0, NULL, 0).
```

Similarly, all sending calls behave the same if they are called as follows:
```
        write(sockfd, buf, len);
or
        send(sockfd, buf, len, 0);
or
        sendto(sockfd, buf, len, 0, NULL, 0).
```

3.3.3   Byte ordering functions

There are some useful functions that are available to convert between network byte order and host byte order. These are defined in the following header file.
#include<netinet/in.h>
The different functions for this purpose are given as follows:

i) Host to network byte order

> `uint32_t htonl(uint32_t hostlong)`
> This converts the long integer hostlong from host byte order to network byte order.
>
> `uint16_t htons(uint16_t hostshort)`
> This converts the short integer hostshort from host byte order to network byte order. Both the above functions return the values in network byte order.

ii) Network to host byte order

> `uint32_t ntohl(uint32_t netlong)`
> This converts the long integer netlong from network byte order to host byte order.
>
> `uint16_t ntohs(uint16_t netshort)`
> This converts the short integer netshort from network byte order to host byte order. Both the above functions return the values in host byte order.

3.3.4   Address conversion functions

These functions are used to convert an IP address between a human readable ASCII string of dotted decimals (like 10.10.10.1 in IPv4) and network byte ordered binary (numeric) values that are stored in a socket address structure.

i) **ASCII to numeric**

```
int inet_aton(const char *strptr,
struct in_addr *addrptr);
```
This converts an ASCII string IPv4 address to numeric address. It the returns 1 if string is a valid IPv4 address and otherwise returns 0.

```
in_addr_t inet_addr(const char *strptr);
```
This converts a dotted decimal ASCII string to a corresponding network byte ordered 32-bit binary value. It returns INADDR_NONE in case of error.

```
int inet_pton(int family, const char
strptr, void *addrptr);
```
This converts an ASCII string (called a presentation) of either an IPv4 or IPv6 address to a corresponding numeric address. It returns 1 if the string is a valid IP address, 0 if a string is not a valid IP address, and −1 on error.

i) **Numeric to ASCII**

```
char* inet_ntoa(struct in_addr inaddr);
```
This converts a numeric IPv4 address to a dotted decimal ASCII string. It returns a pointer to the string.

```
const char *inet_ntop(int family, const void
*addrptr, char *strptr, size_t len);
```
This converts a binary network byte order of either an IPv4 or IPv6 address to a corresponding ASCII string. It returns a pointer to the string.

3.3.5   Functions for protocol addresses

These functions are used to get the IP address and/or port number of either the local machine or a foreign machine. They get these addresses from the kernel space to the user space.

i) **Local Protocol Address**

```
int getsockname(int sockfd, struct sockaddr
*localaddr, socklen_t *addlen);
```

ii) **Foreign Protocol Address**

```
int getpeername(int sockfd, struct sockaddr
*peeraddr, socklen_t *addlen);
```

These functions are required in the following cases:

- In a TCP client when *bind()* is not called, these functions may be used after the *connect()* call to get the IP address and port number assigned.

- In a TCP server when *bind()* is called with port=0 and the IP address is set as a wild card, the *getsockname()* may be used with *sockfd* as a socket descriptor after the *bind()* call to get local addresses with *sockfd*, and *getpeername()* may be used with *connfd* (as a socket descriptor) after the *accept()* call to get foreign addresses.

3.3.6   Functions for hostname

```
#include<unistd.h>
int gethostname(char *name, size_t len);            int
sethostname(const char *name, size_t len);
```
These system calls are used to access or to change the host name of the current processor. *sethostname()* sets the host name to the value given in the character array name. The *len* argument specifies the number of bytes in the name. *gethostname()* returns the null-terminated host name in the character array name, which has a length of *len* bytes. On success, 0 is returned. On error, -1 is returned.

Listing 3.4: Code segment to display host name

```
1 char host[30];
2 gethostname(host, 30);
3 printf("Host Name = %s\n", host);
```

```
#include<netdb.h>
struct hostent *gethostbyname(const char *name);
```

```
#include<sys/socket.h>
struct hostent *gethostbyaddr(const void *addr,
socklen_t len, int type);
```

The *gethostbyname()* function returns a structure of type *hostent* for the given host name. The name is either a hostname or an IPv4 address in standard dot notation. If the name is an IPv4, the *gethostbyaddr()* function returns a structure of type *hostent* for the given host address *addr* of length *len* and address type *type*. The host address argument is a pointer to a structure of a type depending on the address type, for example, a structure *in_addr* * for address type AF_INET.

```
struct hostent {
        char *h_name;/*official name of host*/
        char **h_aliases;/*alias list*/
        int h_addrtype;/*host address type*/
        int h_length;/*length of address*/
```

```
        char **h_addr_list;/*list of addresses*/
};
```

Example:

Listing 3.5: Code segment to display the host name along with list of IP addresses

```
1   struct hostent *host;
2   struct in_addr **addr_list;
3   host=gethostbyname("host_name");
4
5   printf("Name is: %s\n", host->h_name);
6   printf("IP addresses:\n");
7   addr_list = (struct in_addr **)host->h_addr_list;
8
9   for (i = 0; addr_list[i] != NULL; i++) {
10    printf("%s \n", inet_ntoa(*addr_list[i]));
11  }
```

Output:
The hostname and the list of IP addresses for that host.
To check the output values, the OS command *hostname* and *ifconfig* may be used for hostname and IP addresses, respectively.

Test/Viva Questions

a) List the calls used in a TCP server in sequence to establish a connection with the client.

b) Draw a block diagram to represent the structure of a UDP client/server program.

c) What are the parameters for a *bind()* call?

d) What is returned from an *accept()* call?

e) What is the significance of the second argument to a *listen()* call?

f) Why are two sockets needed in a TCP server? How many sockets are required in a UDP server?

g) What happens if a *getsockname()* call is not used in a program?

h) Differentiate between host and network byte order. Which format is used in TCP/IP?

i) How many bits are there to store a port number?

j) Why does a *sendto()* call take more parameters as compared to a *send()* call?

k) What modification in the server program is required to assign a valid IP and a port number to the server socket address structure in place of IN_ADDR_ANY and zero, respectively?

l) What are the parameters used in a *connect()* call?

m) What is the use of the *ntoa*() function?

n) What is returned by the function *gethostbyaddr*()?

o) What is the sequence of calls needed in a UDP client program? How is it different from a TCP client?

Programming Assignments

1) Write any TCP server program that uses fixed IP addresses and port numbers. The client program passes only IP addresses from the command line and fixes the server port number inside the program.

2) Write programs to achieve the following. TCP client and server programs print the sender's IP:port number along with the message received. When a client closes, the server prints this information along with the IP of the client.

3) Write a client program that takes a host name in the command line and connects to that host (server is running in the given host before the client).

4) Write a UDP client/server program as per the block diagram given in Figure 3.4a. The server receives some messages from the client.

5) Write a UDP client/server program in which the UDP server runs forever. Run multiple clients, with each client sending multiple messages to the server. The server echoes back the messages to respective clients. Both client and server print the sender's IP:Port along with the message received. The client stops after sending a "bye" message.

6) Write a TCP client/server program to transfer files between server and client. If the client puts the command put/get <file-name>, then the file is transferred to/from the server accordingly.

7) Write a UDP client/server program to transfer files between server and client. If the client issues the command put/get <file-name>, then the file is transferred to/from the server accordingly.

8) Write a TCP client/server program in which an operating system command is sent from the client to the server. The server executes the command locally and sends back the output of the command to the client.

9) Write a UDP client/server program in which an operating system command is sent from client to server. The server executes the command locally and sends back the output of the command to client.

10) Write a TCP client/server program in which the server contains a function to sort an array of numbers. The client sends an array of

numbers to the server, and the server sorts it using its own function and returns the result to the client. The client prints the sorted numbers.

3.4   I/O Multiplexing

Most of the calls discussed so far use blocking calls such as *accept()* or *recv()*, i.e., if a *recv()* call is placed but no data are available, then it blocks until data become available and copied to the receive buffer. The situation is demonstrated in figure 3.10. During this blocking period, the program is unable to handle any other descriptor that might be ready by this time. For example, a close request may be available in a socket descriptor, but the process may not know this if it has already blocked reading from standard input. There are many models available to overcome this situation, such as a *nonblocking model, I/O multiplexing model, signal driven I/O model*, and *asynchronous model*, etc. Here we will discuss the I/O multiplexing model, which is the most effective and widely used model. However, readers are encouraged to consult the references provided in the Bibliography for other models.

The I/O multiplexing model notifies the program when one or more I/O are ready. In simple words, I/O multiplexing is required when either a client or a server is handling multiple descriptors like a listening socket, connected sockets, and/or standard input, etc. To implement I/O multiplexing, the *se-*

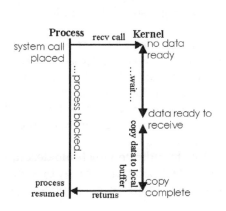

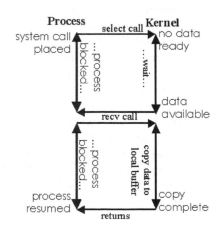

Figure 3.10: Sequence of actions performed in the case of a blocking I/O.

Figure 3.11: Sequence of actions performed in the case of I/O multiplexing.

lect() or *poll()* system call is used. The technique is that now the process blocks on these calls, unlike blocking I/O, where the process blocks in the actual I/O call. The time line diagram for both the blocking and I/O multiplexing cases is provided in Figures 3.10 and 3.11. As can be seen from the figures, in the case of the blocking I/O model, the process blocks in *recv()* call, whereas in the case of I/O multiplexing, the process blocks in the select call, not in the *recv()* call. Looking at these figures, one is not very encouraged, as in both cases the process blocks, i.e., in *recv()* in the former case or in *select()* in the latter case. Also some disadvantages may be noticed in the latter case in that it uses two system calls to perform the same action (recv in this case) as opposed to one system call in the former case. But actually the latter has an tremendous advantage over the former one only because of the fact that *select()* can wait for more than one descriptor to be ready, whereas the blocking model can wait for only one descriptor to be ready for which the call is placed. This idea will be more understandable when we walk through several examples that follow in this section.

3.4.1   Synchronous I/O multiplexing using select() call

The prototype for *select()* call is as follows.

```
<sys/select.h>
int select(int nfds, fd_set *readfds, fd_set *writefds,
fd_set *ecceptfds, struct timeval *timeout);
```

The *select()* call allows a process to monitor multiple file descriptors, waiting until one or more of the file descriptors become "ready" for some type of I/O operation. A file descriptor is considered ready if it is possible to perform the corresponding I/O operation (e.g., *read, recv,* etc.) without blocking.

The first argument, *nfds*, represents the highest-numbered file descriptor in any of the three sets, plus 1.

The next three arguments consist of three different sets of file descriptors that are watched. Those listed in *readfds* will be watched to see if characters become available for reading, i.e., to see if a read will not block; in particular, a file descriptor is also ready on the end-of-file; those in *writefds* will be watched to see if a write will not block, and those in *exceptfds* will be watched for exceptions. If you don't want to watch all types of descriptors at the same time, then some of the lists may be set to *NULL*.

On exit, the sets are modified in place to indicate which file descriptors actually changed status. Each of the three file descriptor sets may be specified as NULL if no file descriptors are to be watched for the corresponding class of events.

The last argument, *timeout*, is an upper bound on the amount of time that elapses before *select()* returns. It is expected that *select()* returns before timeout time. The *timeval* structure is defined in <sys/time.h> as follows.

```
struct timeval {
        long tv_sec;/* seconds */
        long tv_usec;/* microseconds */
};
```

Also, when the function returns, timeout is updated with time still remaining. If time is set to 0, then select returns immediatly. If it is set to *NULL*, then it blocks until any of the available descriptors become ready.

Four macros are available to manipulate the descriptor sets as follows.

i) FD_ZERO(fd_set *fdset)   clears all entries from the set

ii) FD_SET(int fd, fd_set *fdset)   adds a given file descriptor to the set

iii) FD_CLR(int fd, fd_set *fdset)   removes a given file descriptor from a set

iv) FD_ISSET(int fd, fd_set *fdset)   returns *true* if file descriptor *fd* is part of the set. Normally used after *select()*.

On success, *select()* returns the *number* of file descriptors contained in the three returned descriptor sets. The value may be *zero* if the timeout expires before any descriptor is ready. On error, -1 is returned, and the sets and timeout become undefined, so do not rely on their contents after an error.

Now it is time to actually use the *select()* call in a program to test its behavior. Look at the following program in Listing 3.6 that checks to see if there is some data from standard input (keyboard) within a specified time interval.

Listing 3.6: A simple program using select()

```
1  #include <stdio.h>
2  #include <stdlib.h>
3  #include <sys/select.h>
4
5  int main(void){
6    fd_set readfds;  /*descriptor list.*/
7      struct timeval tv;
8      int res;
9
10    FD_ZERO(&readfds); /*clear the set entries.*/
11    FD_SET(0, &readfds);  /*add stdin descriptor to set.*/
12
13      tv.tv_sec = 5;
14      tv.tv_usec = 500000; /*Wait up to 5.5 seconds.*/
15
16      res = select(1, &readfds, NULL, NULL, &tv);
17
18      if(res == -1)
19            perror("select()");
20      else if(res)  /* FD_ISSET(0, &readfds) will also be true.*/
21            printf("Data available with stdin.\n");
22      else
23            printf("No data in stdin within 5.5 seconds.\n");
24  }
```

When Listing 3.6 is executed, it blocks in select; if something is entered from the keyboard within 5.5 sec, it returns and line 21 is executed. If no data entered within 5.5 secs, it returns and line 23 is executed. If error occurs, Line 19 is executed, after which the program ends.

Now let us modify the above program such that the select call blocks until data is entered from the keyboard (unrestricted time). When data is entered from the keyboard, select returns and the data is copied to the process buffer and displayed. This process continues until a bye message is entered from the keyboard. This program is given in Listing 3.7.

Listing 3.7: Reading data after select returns

```
1  #include <stdio.h>
2  #include <stdlib.h>
3  #include <string.h>
4  #include <sys/select.h>
5
6  int main(void){
7      fd_set readfds;
8          char buf[1024];
9
10     FD_ZERO(&readfds);/*clear the set entries.*/
11     FD_SET(0, &readfds); /*add stdin descriptor to set.*/
12
13     do{
14             printf("Enter a message from keyboard(bye to quit):");
15         if(select(1, &readfds, NULL, NULL, NULL)==-1){
16             perror("select()");
17             exit(1);
18             }
19     memset(buf,0,1024);
20         if(FD_ISSET(0, &readfds)){ /*stdin descriptor ready*/
21             read(0, buf, sizeof buf); /*copy data to buffer*/
22             printf("Message Typed: %s\n",buf); /*display*/
23             }
24     }while(strcmp(buf, "bye\n") != 0);
25  }
```

select() returns when the stdin is ready in line 15, which is checked in line 19, copied to the local buffer in line 20, and then displayed in line 21. The work done in Listing 3.7 is very simple and the same result can be obtained using a single *scanf()* function (this also blocks) without using the *select()* and *read()* calls. Actually, the concern is not with the output, but with the way select can be used. The real application can be visualized when more than one descriptor is watched as opposed to one descriptor (stdin) watching in this program.

To consider a more realistic situation, let us consider a situation where there is a communication between one client and one server. This program was discussed in the previous section, but in that case, if a *recv()* call is used in a process, then the process cannot send any message to the other process until *recv()* is finished. That is, *send()* and *recv()* calls cannot be used randomly by any communicating party, i.e., they have to follow a sequence of alternating *send()* and *recv()* calls; more than one consecutive *send()* call from one party is not possible, which is a practical requirement, when two persons are chatting with each other remotely. The program given in Listing 3.8 demonstrates two persons chatting where one can *send()* or *recv()* randomly, there can be more than one *send()* call executed by one party without executing any *recv()* call.

Listing 3.8: Select server

```
1  #include<stdio.h>
2  #include<string.h>
3  #include<stdlib.h>
4  #include<netinet/in.h>
5  #include<sys/select.h>
6
7  int main(){
8    int fd, tempfd, sockfd, confd = -1;
9    char buf[1024];
10   struct sockaddr_in server, client;
11   socklen_t cliAddrLen;
12   int result, fdmax, nread, flag = 0;
13   fd_set readfds, finfds;
14
15   sockfd=socket(AF_INET, SOCK_STREAM, 0);
16   if(sockfd < 0){
17     perror("\nSocket()");
18     exit(1);
19   }
20
21   server.sin_family = AF_INET;
22   server.sin_addr.s_addr = htons(INADDR_ANY);
23   server.sin_port = htons(4321);
24   if(bind(sockfd, (struct sockaddr*)&server, sizeof(server))){
25     perror("\nBind()");
26     exit(2);
27   }
28       printf("\nSocket has port # %hd \n", ntohs(server.sin_port
       ));
29
30   listen(sockfd, 5);
31       FD_ZERO(&finfds);
32   FD_ZERO(&readfds);
33
34   /*Add stdin and sockfd to final descriptor list.*/
35   FD_SET(0, &finfds);
36   FD_SET(sockfd, &finfds);
37   fdmax = sockfd; // descriptors# in the read descriptor list
38
39   while(1){//continue for ever
40     readfds = finfds;//copy final list to read descriptor list
41     /*block on select till an activity found*/
42     result = select(fdmax+1, &readfds, NULL, NULL, NULL);
43     if(result < 1){
44       perror("Select()");
45       exit(3);
46     }
47
48     for(fd=0; fd <= fdmax; fd++){//For all recorded fds
49       if(FD_ISSET(fd, &readfds)){//Activity found on which fd?
```

```
50        if(fd == 0){//if written from server keyboard
51           memset(buf, 0, sizeof buf);
52           read(0, buf, 1024);//read keyboard
53           if(flag){//send to the client if exists
54              send(confd, buf, sizeof buf, 0);
55           }
56           else{//if no client exists
57              printf("First connect from a client...\n");
58           }
59        }//endif read from keyboard
60        else if(fd == sockfd){
61           //If a connection request comes from client.
62           if(confd > 0){
63              //If connection exists, do not allow more
connections.
64              cliAddrLen = sizeof(client);
65              tempfd=accept(sockfd, (struct sockaddr*)&client, &
cliAddrLen);
66              memset(buf, 0, sizeof buf);
67              sprintf(buf,"%s","Already Talking with somebody,
Please Wait...\n");
68              send(tempfd, buf, sizeof buf, 0);
69              close(tempfd);
70           }
71           else{
72              //If no existing connection available, then connect
to it.
73              cliAddrLen = sizeof(client);
74              confd = accept(sockfd, (struct sockaddr *)&client, &
cliAddrLen);
75              if(confd < 0){
76                 perror("Accept");
77                 exit(3);
78              }
79              FD_SET(confd, &finfds);
80              fdmax = confd;
81              flag = 1;
82
83                      printf("Client connected on confd: %d\n",
confd);
84           }
85        }//endif accept
86        else{//if it is not sockfd then it must be confd
87           memset(buf, 0, sizeof buf);
88           if((nread = recv(fd, buf, sizeof buf, 0)) <= 0){
89              //if connection closed from client
90              printf("Socket %d closed\n", fd);
91              printf("Removing client on fd: %d\n", fd);
92              close(fd);
93              FD_CLR(fd, &finfds);
94              confd = -1;
95              flag = 0;
96           }
97           else{//Print the received message
98              printf("Client: %s\n", buf);
99           }
```

```
100        }//endif recv
101       }//endif activity found
102     }//for loop ends here
103    }//while loop ends here
104 }//main ends
```

Lines 8 to 13 are used to declare/initialize different variables used in the program. Lines 15 to 30 are the usual calls for a stream (TCP) server with a socket ready to listen for incoming requests. Lines 31 to 32 are used to clear the final descriptor set and read descriptor set. As a select call modifies *readfds*, therefore another *finfds* is used to keep actual descriptor sets. Lines 35 to 36 add two descriptors, the *stdin*, used to type a message from the keyboard to send to the client, and the listening socket *sockfd*. Line 37 sets the value of *fdmax* to the highest descriptor number so far, i.e., *sockfd*. Line 39 starts a infinite loop for a continuous communication between two processes. Lines 42 to 46 call *select()* along with error checking. The call updates *reafds* with an active descriptor set. Lines 48 to 49 find the ready descriptors out of recorded descriptors and take action as required by different active descriptors in the following lines of codes.

Lines 50 to 59 deal with a message typed from a server keyboard. They reads the message to the local buffer (*buf*), then check if a connection is already available using a flag. If a flag is true, then it sends it to the other end process (line 54); otherwise it skips the message sending by printing a warning message in line 57.

Lines 60 to 85 deal with an incoming connection request from a client process. Because we have assumed only a two-way communication in this program, more than one connect request is rejected in lines 62 to 70. Otherwise the connection is accepted in lines 71 to 78. When it is connected, the connected descriptor is created and the new descriptor is added to the *finfds* set in line 79. In lines 80 to 81 the *fdmax* is set to a new value and the connection flag is set to true, i.e., no new connection is allowed.

Finally, lines 86 to 99 deal with a message arriving from the client. Lines 88 to 96 handle a connection close request from the client. In this case the descriptor needs to be removed from the *finfds* set, which is done in line 93; *confd* is set to negative number in line 94 to represent a no connection available state, and a connection flag is set to false in line 95. If it is not a connection close request, then it must be a message from the client, which is printed in line 98. The rest of the lines are closing scopes for corresponding opening scopes in the program.

Now let us understand the output produced. First compile the program and execute it as shown in lines 1 and 2 of Figure 3.12. It will print the port number (4321 in this case). Then open another terminal and connect the server with a telnet client with following command.

```
telnet <server_ip> <server_port>
```

as shown in Figure 3.13. If connected successfully, a message is printed in the server terminal mentioning the descriptor number. Then normal communi-

```
ajit@ajit-HP-ProBook-4420s: ~/bookProg/chap2
ajit@ajit-HP-ProBook-4420s: ~/bookProg/ch...  ✖  ajit@ajit-HP-ProBook-4420s:~/bookProg/ch...  ✖  ajit@ajit-HP-ProBook-4420s:~/t
ajit@ajit-HP-ProBook-4420s:~/bookProg/chap2$ cc selServer1.c -o server
ajit@ajit-HP-ProBook-4420s:~/bookProg/chap2$ ./server

Socket has port # 4321
Client connected on confd: 4
Client: Helo

Hi
Client: how are you?

Client: who are you?

I am fine
I am Ajit
Socket 4 closed
Removing client on fd: 4
Client connected on confd: 4
Client: hi there

```

Figure 3.12: Select server terminal

```
ajit@ajit-HP-ProBook-4420s:~/bookProg/chap2
ajit@ajit-HP-ProBook-4420s:~/bookProg/ch...  ✖  ajit@ajit-HP-ProBook-4420s:~/bookProg/ch...  ✖  ajit@ajit-HP-ProBook-44
ajit@ajit-HP-ProBook-4420s:~/bookProg/chap2$ telnet 127.0.0.1 4321
Trying 127.0.0.1...
Connected to 127.0.0.1.
Escape character is '^]'.
Helo
Hi
how are you?
who are you?
I am fine
I am Ajit
^]

telnet> quit
Connection closed.
ajit@ajit-HP-ProBook-4420s:~/bookProg/chap2$
```

```
ajit@ajit-HP-ProBook-4420s:~/bookProg/chap2
ajit@ajit-HP-ProBook-4420s:~/bookProg/ch...  ✖  ajit@ajit-HP-ProBook-4420s:~/bookProg/ch...  ✖  ajit@ajit-HP-ProBook-442
ajit@ajit-HP-ProBook-4420s:~/bookProg/chap2$ telnet 127.0.0.1 4321
Trying 127.0.0.1...
Connected to 127.0.0.1.
Escape character is '^]'.
Already Talking with somebody, Please Wait...
Connection closed by foreign host.
ajit@ajit-HP-ProBook-4420s:~/bookProg/chap2$ telnet 127.0.0.1 4321
Trying 127.0.0.1...
Connected to 127.0.0.1.
Escape character is '^]'.
hi there
```

Figure 3.13: First client terminal　　　Figure 3.14: Second client terminal

cation goes on and anybody can send at any time, i.e., one can send multiple messages without receiving any message, unlike the programs not using *select()*, where, after sending one message, it has to receive one message from the other party before sending the second message.

Now open the third terminal and try to connect to the same server, as shown in Figure 3.14. It receives a message from the server that one connection is already in place and it rejects the request by closing it. Now close the first client by pressing *ctrl+]* followed by a *quit* command, as shown in Figure 3.13. The connection close message is printed in the client server and the server prints the removal of the descriptor. Now if you try to connect from the second client, this will be accepted. At any time only one client can be connected. One more practical use of the select could be to design a multi-client chat server. Unlike the above program, the server here does not take part in actual communication, but facilitates communication between other clients. The number of clients can be as many as desired. The server receives a message from any client and forwards this message to all other clients connected at that time, excluding the sending client and the server itself. The code is given in Listing 3.9.

Listing 3.9: Multiclient chat server

```
1  #include<stdio.h>
2  #include<string.h>
3  #include<stdlib.h>
4  #include<netinet/in.h>
5  #include<sys/select.h>
6
7  int main(){
8    int fd, sockfd, confd;
9    int i, fdmax, nread, flag = 0;
10   char buf[1024];
11   struct sockaddr_in server, client;
12   socklen_t cliAddrLen;
13   int result;
14   fd_set readfds, finfds;
15
16   sockfd = socket(AF_INET, SOCK_STREAM, 0);
17   if(sockfd < 0){
18     perror("\nSocket Open... ");
19       exit(1);
20   }
21
22        server.sin_family = AF_INET;
23     server.sin_addr.s_addr = htons(INADDR_ANY);
24     server.sin_port = htons(6666);
25     if(bind(sockfd, (struct sockaddr *)&server, sizeof(server)))
        {
26       perror("\nBind ... ");
27       exit(2);
28   }
29
30   printf("\nSocket has port # %hd \n", ntohs(server.sin_port));
31     listen(sockfd, 5);
32
33     FD_ZERO(&readfds);
34     //Add sockfd
35     FD_SET(sockfd,&finfds);
36     fdmax = sockfd; //max number of descriptors
37
38   while(1){//loop for ever
39        readfds = finfds;
40        result = select(fdmax+1, &readfds, NULL, NULL, NULL);
41        if(result < 1){
42          perror("Select:");
43          exit(1);
44        }
45
46     for(fd = 0; fd <= fdmax; fd++){
47       if(FD_ISSET(fd, &readfds)){//if activity found
48         if(fd == sockfd){//if connection request arrives
```

```
49            cliAddrLen = sizeof(client);
50            confd = accept(sockfd, (struct sockaddr *)&client, &
       cliAddrLen);
51            if(confd < 0){
52               perror("Accept");
53               exit(3);
54            }
55            FD_SET(confd, &finfds);//attach confd to descriptor
       list.
56          if (confd > fdmax) {     // keep track of the max
57             fdmax = confd;
58          }
59          printf("Adding client on confd: %d\n", confd);
60        }//endif conn request
61        else{//if data arrives
62        memset(buf, 0, sizeof buf);
63           if((nread = recv(fd, buf, sizeof buf, 0)) <= 0){//if
       client closes
64             printf("socket %d closed\n", fd);
65             printf("Removing client on fd %d\n", fd);
66             close(fd);
67             FD_CLR(fd, &finfds);
68           }//endif client closing
69           else{//send the message to all other client.
70           for(i = 0;i <= fdmax; i++){
71             if(FD_ISSET(i, &finfds)){
72                if(i != sockfd && i != fd)
73                   //donot send to server or sending client
74                   send(i, buf, sizeof buf, 0);
75             }
76           } //all client finished
77         }//sending done
78        }//endif data arrival
79      }
80    }//for loop end
81    }//while loop end
82 }//main ends
```

The logic for Listings 3.9 and 3.8 are the same. The differences in the current program are

i) The current program can connect from any number of clients, unlike the program in Listing 3.8, which can connect to only one client. That is, lines 62 to 70, of Listing 3.8 are absent in the current program.

ii) The *stdin* descriptor is not used because the server is not going to send its own message but only forwards the message received from a client to all other clients, which is expressed in lines 69 to 77 of Listing 3.9.

The program execution demonstration is shown in Figure 3.15 to Figure 3.18. Figure 3.15 represents the server terminal and the other three figures represent three different client terminals connected simultaneously to the server. As can be seen from the figures, any message sent from any client is forwarded to the remaining clients by the server.

```
ajit@ajit-HP-ProBook-4420s:~/bookProg/chap2$ cc selServer2.c -o server
ajit@ajit-HP-ProBook-4420s:~/bookProg/chap2$ ./server

Socket has port # 6666
Adding client on confd: 4
Adding client on confd: 5
Adding client on confd: 6
socket 6 closed
Removing client on fd 6
socket 4 closed
Removing client on fd 4
socket 5 closed
Removing client on fd 5
```

Figure 3.15: Multiclient chat server

```
ajit@ajit-HP-ProBook-4420s:~/bookProg/chap2$ telnet 127.0.0.1 6666
Trying 127.0.0.1...
Connected to 127.0.0.1.
Escape character is '^]'.
Can you here me?
I am Ajit
ok I am Ashok
hey its me
who are there?
^]

telnet> quit
Connection closed.
ajit@ajit-HP-ProBook-4420s:~/bookProg/chap2$
```

Figure 3.16: First client terminal

```
ajit@ajit-HP-ProBook-4420s:~/bookProg/chap2$ telnet 127.0.0.1 6666
Trying 127.0.0.1...
Connected to 127.0.0.1.
Escape character is '^]'.
Can you here me?
I am Ajit
ok I am Ashok
hey its me
who are there?
Any body there?
^]

telnet> quit
Connection closed.
ajit@ajit-HP-ProBook-4420s:~/bookProg/chap2$
```

Figure 3.17: Second client terminal

```
ajit@ajit-HP-ProBook-4420s:~/bookProg/chap2$ telnet 127.0.0.1 6666
Trying 127.0.0.1...
Connected to 127.0.0.1.
Escape character is '^]'.
Can you here me?
I am Ajit
ok I am Ashok
hey its me
who are there?
Any body there?
^]

telnet> quit
Connection closed.
ajit@ajit-HP-ProBook-4420s:~/bookProg/chap2$
```

Figure 3.18: Third client terminal

Test/Viva Questions

a) List the limitations of blocking calls of the socket API.

b) List some blocking calls of the socket API.

c) What are the different models available to overcome the limitations of blocking I/O? Which is the most acceptable model?

d) What is I/O multiplexing? Give an example.

e) How is the I/O multiplexing model different from the blocking I/O?

f) What is the prototype for select call?

g) What is the datatype and use of the timeout parameter used in the *select()* call?

h) What is the macro used to add a file descriptor to the set?

i) What is the use of the macro *FD_ISSET*?

j) What is returned from a select call?

k) Write separate statements to assign timeout values for: 2.8 sec, 100 ms, and 5.005 sec.

Programming Assignments

1) Write a program using the *select()* call that captures and prints any data entered from the keyboard within 2.3 sec.

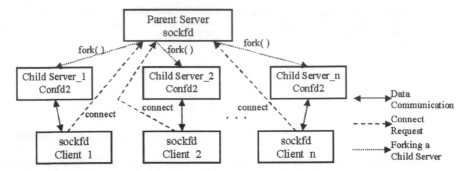

Figure 3.19: Block diagram of concurrent connections

2) The server program given in Listing 3.8 uses a telnet client. Develop a client program that works in a similar way for that server.

3) Modify the server program given in Listing 3.8 in which the server prints the IP:port of a client that is currently connected/closed.

4) Modify the server program given in Listing 3.8 such that the server echoes back the respective messages to respective clients.

5) Modify the server program given in Listing 3.8, to allow connections from three clients only. Each client requests a different file from the server. The server sends appropriate files to the corresponding clients.

3.5   Concurrency

In the case of I/O multiplexing, we observe that a server can connect to multiple clients simultaneously and *send* and *recv* can be performed randomly without blocking. But consider another situation where each client needs different information from the server. For example, in the case of a multiclient remote login server, if each client executes a different command, then the response needs to be different, corresponding to the individual client's request. Similarly, in the case of a file server, individual clients may request different files for upload or download, and the response from the server must be in accordance with the request imposed by the client. It becomes difficult to handle this kind of scenario using I/O multiplexing only.

A better solution could be that, for each connection request from a client, the server creates a child server process to handle that client. The lifetime of these server children ceases with the closing of respective clients. Each server child takes care of a single client at a time. This concept is depicted in Figure 3.19

As given in Figure 3.19, consider a connection request from client 1 (dashed line) to the server. The server uses its listening socket (*sockfd*) to take on the request. The server then forks a child server_1 process that accepts the connect request of the client 1 process. The connected socket is created in child server_1 process so that the child server_1 will now be responsible for communication with the client 1 process, and the parent server process remains free to accept requests from some other client. In the mean time, say client 2 issues a connection request to the parent server; then the above steps are repeated to create another child server_2 process to handle communication with the client 2 process, and the parent server process becomes free again to listen to connection requests from any other client process. This process goes on forever. That is, in this technique the server process is able to handle multiple clients simultaneously by creating a separate child server process for each client.

The *fork()* system call is used to create a child server process. Let us now understand the syntax and behavior of a *fork()* system call.

fork() system call

- *fork()* is used to create a duplicate of a process called the child process.

- *fork()* takes no arguments, returns an integer

- if *fork()* is called once, it returns twice

  - Once in the calling process called the parent process, with an integer value called the *ppid* (parent process ID)

  - Once in the newly created process called the child process, with value 0

- The child may call *getppid()* to get the parent process ID

- The child is an exact copy of the parent process.

- All descriptors opened in the parent are shared by the child, and the program execution resumes from the statement after the fork call.

Listing 3.10 is intended to illustrate the simple use of a fork call.

Listing 3.10: A simple program using fork

```
1 #include<stdio.h>
2
3 int main(){
4    fork();
5    printf("\n Hello How are you? \n");
6 }
```

When executed, the program prints the string constant "Hello How are you?" two times. If, in place of one fork call, we use two calls, then it will print the string four times, and so on. It is also possible that, after the creation of the child process, it is required that the parent and child processes perform different tasks, as given in Listing 3.11.

Listing 3.11: Different tasks in parent and child process

```
1  #include<stdio.h>
2
3  int main(){
4    int pid;
5    pid = fork();
6
7    if (pid == 0) //if child process
8      printf("\n I am the child and my PID is: %d \n", getpid());
9    else //if parent process
10     printf("\n I am the Parent and my PID is: %d \n", getpid());
11   }
```

In Listing 3.11, after a child is created in line 5, there are two copies of the same process available, in which the *pid* value 0 is returned to the child process and *ppid* to the parent process. Therefore line 8 is executed in the child process only, and line 10 is executed in the parent process only. Obviously, the *getpid()* function will return different values in each process. In other words, the if block contains the code to be executed only in the child process, whereas the else block will be executed by the parent only.

This technique is used in the case of concurrent server design. The steps followed are as given below.

i) Create a stream socket.

ii) Bind it with a local protocol address.

iii) Convert the passive socket into a listening socket.

iv) If a connection request arrives from a client

    a) Accept the connection.

    b) Fork a process.

    c1) Tasks for the child process

        1) Close the listening descriptor (only the server can handle a connection request).

        2) Talk with the client (using *confd*).

        3) Repeat step 2 until one side closes.

        4) Close the connected descriptor when finished.

        5) Exit the child server (the child server is no longer required).

    c2) Task for the parent process

        • Close the connected descriptor (the parent does not talk with a client).

v) Repeat step *iv* above forever

The above algorithm is implemented as given in Listing 3.12. The code for the client program is given in Listing 3.13.

Listing 3.12: Concurrent server program

```
1  #include<sys/socket.h>
2  #include<netinet/in.h>
3  #include<netdb.h>
4  #include<stdio.h>
5  #include<string.h>
6  #include<signal.h>
7  #include<stdlib.h>
8  #define TRUE 1
9
10 int main(){
11   int sockfd, confd, chpid, clCnt=0;
12   char buf[1024];
13   struct sockaddr_in server, client;
14   socklen_t cliAddrLen = sizeof(client), servAddrLen = sizeof(
     server);
15
16   sockfd = socket(AF_INET, SOCK_STREAM, 0);
17   if(sockfd < 0){
18     perror("\n Socket ... ");
19     exit(1);
20   }
21
22   server.sin_family = AF_INET;
23   server.sin_addr.s_addr = htons(INADDR_ANY);
24   server.sin_port = 0;
25   if(bind(sockfd,(struct sockaddr *)&server, sizeof(server))){
26     perror("\n Bind ... ");
27     exit(2);
28   }
29
30       if(getsockname(sockfd, (struct sockaddr *)&server, &
     servAddrLen)){
31     perror("\n Getting port... ");
32     exit(3);
33   }
34
35   printf("\n Socket has port # %d \n ", ntohs(server.sin_port));
36   listen(sockfd, 5);
37
38   do {
39     /* handle a request from a client*/
40     confd = accept(sockfd, (struct sockaddr *)&client, &
     cliAddrLen);
41     if(confd == -1) {
42       perror("\n Accept... ");
43       exit(4);
44     }
45     else {
46       clCnt++;
```

```
47      if((chpid = fork()) == 0) {    //create the child server
48        close(sockfd);/*close the listening socket in child*/
49        printf("\n \t ---child Server_%d created---\n ",clCnt);
50
51        do{      //serv the client
52          memset(buf, 0, sizeof buf);
53          recv(confd,buf, 1024, 0);
54          printf("\n Message from client_%d :: %s\n ", clCnt,
      buf);
55        }while(strcmp(buf,"bye")!=0);
56
57            close(confd); //close connected socket
58        printf("Client_%d closed and hence the child Server_%d\n
      ", clCnt,clCnt);
59        exit(0); //terminate child server
60      }
61      else{
62        close(confd);//close the connected socket in parent
63      }
64
65    }
66  }while(TRUE); //continue for ever
67
68  close(sockfd); // close the parent
69  return 0;
70 }
```

If you look at the server program, lines 1 to 40 correspond to steps *i* to *iv.a* given above and are like any stream server program. If *accept()* returns successfully in line 44, then a child is forked in line 47 that corresponds to step *iv.b*. The code segment given in lines 47 to 60 is executed by the newly created child server where the child closes the listening socket in line 48 and serves the client until client sends a "bye" message in a do-while loop given in lines 51 to 55. At the same time the parent process closes the connected socket in lines 61 to 63 as these lines correspond to the parent server process and returns to line 51 and waits for further incoming requests from a different client. If again a connection request comes from another client, the same fork is repeated by the parent to create the second child server, and so on. Parent and children are executed in parallel and perform their respective tasks separately and concurrently. When any child server finishes serving a corresponding client, the child is no longer required and hence the child's connected socket is closed in line 62, after which that child server is terminated in line 69 and the life of that child ends. But some other children may be running at the same time serving other clients. The same parent program runs forever, creating and destroying the children as and when required, and all the children may run concurrently, serving multiple clients simultaneously.

The client program given in Listing 3.13 is a simple stream client, as given in Listing 3.3. The only difference is that this client can send multiple messages to a server in one go, whereas the previous one closes after sending one message to the server.

Listing 3.13: Client program

```
1  #include<sys/socket.h>
2  #include<netinet/in.h>
3  #include<netdb.h>
4  #include<stdio.h>
5  #include<string.h>
6  #include<stdlib.h>
7  #include<arpa/inet.h>
8
9  int main(int argc,char *argv[]){
10   int sockfd,saddrlen;
11   char buf[1024];
12   struct sockaddr_in server;
13
14   sockfd=socket(AF_INET, SOCK_STREAM, 0);
15   if(sockfd < 0){
16     perror("\n Socket ... ");
17     exit(1);
18   }
19
20   server.sin_family = AF_INET;
21   server.sin_addr.s_addr = inet_addr(argv[1]);
22   server.sin_port = htons(atoi(argv[2]));
23
24   if(connect(sockfd,(struct sockaddr *)&server, sizeof(server))
       < 0){
25     perror("\n Connection ... ");
26     exit(2);
27   }
28
29   do{
30     printf("\n Client input : ");
31     scanf(" %[^\n]", buf);
32     send(sockfd,buf,sizeof buf,0);
33   }while(strcmp(buf, "bye") != 0);
34
35   close(sockfd);
36   return 0;
37 }
```

The program execution is demonstrated in Figures 3.20 to 3.22. Figure 3.20 shows the server terminal and the other figures show the execution of different clients. As can be seen, when a client connects to the server, the server creates a child server process to handle that client and the parent server remains ready to receive any other clients, connection requests. Similarly, when client 2 and client 3 are connected, the corresponding child servers are created. Now each client can communicate with the server simultaneously. The sequence of execution can be seen from the figures. When a client disconnects ("bye" message), the child handling that client terminates. This server program can handle any number of clients.

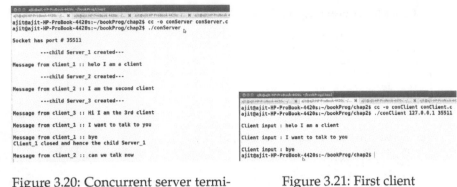

Figure 3.20: Concurrent server terminal

Figure 3.21: First client

Figure 3.22: Second client   Figure 3.23: Third client

Test/Viva/Oral Questions

a) What is a concurrent server? How does it differ from an iterative
   server?

b) How many values are returned from a *fork()* call? What is the significance of these returned values?

c) If a program contains *n* number of *fork()* calls before a *printf()* function, then how many times will the *printf()* function actually be
   executed?

d) If a program contains *n* number of *fork()* calls after a *printf()* function, then how many times will the *printf()* function actually be
   executed?

e) Which function is used to get process id (pid) of a parent process?

f) In a program using a *fork()* call, how do we restrict the portion of
   the program to be executed only by the child process?

g) In Listing 3.12, justify the *exit*(0) call used in line 51.

Programming Assignments

1) Modify the concurrent server program given in Listing 3.12 to restrict the number of concurrent connections to three clients only.

2) Modify the code in Listing 3.12 to echo the messages back to the
   respective clients.

3) Modify the code in Listing 3.12 to use I/O multiplexing in child servers.

4) Write concurrent server/client code that can send different files requested by different clients. It should send an error message if the file is not found.

5) Write concurrent server code having four functions for addition, subtraction, multiplication, and division of two numbers. When a client sends two numbers and an operation, the server executes the corresponding function and sends back the result to the client.

3.6   Broadcasting

Sending a message to all receivers may be achieved by sending the same message iteratively to each receiver one at a time, i.e., multiple transmissions. But sending a message to all receivers simultaneously (with one transmission) can be achieved with UDP and a mechanism called broadcasting. The broadcast message is sent to either of the following addresses.
i) Send the message to subnet's broadcast address (use the *ifconfig* command to find this address).
ii) Send the message to the global broadcast address, 255.255.255.255, which is defined as a constant value called *INADDR_BROADCAST* in the socket API.

However, the message broadcast cannot pass through a router (or a layer-3 switch) from one subnet to another, as broadcasting in the Internet is not allowed, i.e., when a broadcast message reaches a router, it does not forward it to other subnets but drops the packet. Therefore, for both of the above addresses, the message can be received by the hosts available in the same subnet only.
    Also, to use a broadcast, the socket option must be set to *SO_BROADCAST*. Therefore the broadcast program is a UDP program with two modifications. First, set the socket option to *SO_BROADCAST*. The second modification is to send the packet not to a specific destination but to the either of the broadcast addresses given above.
    The code for a broadcast sender (client) is given in Listing 3.14.

Listing 3.14: Broadcast message sender program

```
1  #include <stdio.h>
2  #include <stdlib.h>
3  #include <unistd.h>
4  #include <errno.h>
5  #include <string.h>
6  #include <sys/types.h>
7  #include <sys/socket.h>
8  #include <netinet/in.h>
```

```c
 9  #include <arpa/inet.h>
10  #include <netdb.h>
11  #define  SERVERPORT  7654  // the port used to send
12
13  int main(int argc, char *argv[]) {
14    int sockfd;
15    char msg[1024];
16    struct sockaddr_in bCastListner;
17    int broadcast = 1;
18
19    if (argc != 2) {
20      fprintf(stderr,"usage: ./bcastSender <broadcast IP>\n");
21      exit(1);
22    }
23
24    printf("Enter a Broadcast message:");
25    scanf("%[^\n]", msg);
26
27    if ((sockfd = socket(AF_INET, SOCK_DGRAM, 0)) == -1) {
28      perror("socket");
29      exit(1);
30    }
31
32        // Broadcasting option for socket is set
33    if (setsockopt(sockfd, SOL_SOCKET, SO_BROADCAST, &broadcast,
       sizeof broadcast) == -1) {
34      perror("setsockopt (SO_BROADCAST)");
35      exit(1);
36    }
37
38        bCastListner.sin_family = AF_INET;
39    bCastListner.sin_port = htons(SERVERPORT);
40    bCastListner.sin_addr.s_addr = inet_addr(argv[1]);
41    sendto(sockfd, msg, strlen(msg)+1, 0,  (struct sockaddr *)&
       bCastListner, sizeof bCastListner);
42    printf("sent to %s\n", inet_ntoa(bCastListner.sin_addr));
43    close(sockfd);
44    return 0;
45  }
46  /* Compile with: cc -o bcastSender broadcaster.c */
```

The program in Listing 3.14 is used to send a broadcast message. This is nothing but a simple UDP client program with minor modification. In line 27 the *SOCK_DGRAM* option is passed in the *socket()* function to create a UDP socket. In line 33, the socket option is set to *SO_BROADCAST* to allow broadcasting. Lines 38 to 40 are used to initialize the listener's socket address. In line 41, the message read in lines 24 to 25 is broadcast to the network. Any listeners running at that time can receive the message. Remember that the *sendto()* call does not block, unlike the *send()* call used in TCP sockets, i.e., *sendto()* returns without waiting for a *recvfrom()* to execute. The program stops execution after it by closing the socket.

The code for a broadcast listener (server) is given in Listing 3.15.

Listing 3.15: Broadcast Message Receiver Program

```
1  #include <stdio.h>
2  #include <stdlib.h>
3  #include <unistd.h>
4  #include <errno.h>
5  #include <string.h>
6  #include <sys/types.h>
7  #include <sys/socket.h>
8  #include <netinet/in.h>
9  #include <arpa/inet.h>
10 #include <netdb.h>
11 #define MYPORT   7654 // broadcaster sends to this port
12 #define MAXBUFLEN 1024
13
14 int main(void) {
15   int sockfd, numBytes;
16   char buf[MAXBUFLEN], sendAdd[100];
17   struct sockaddr_in listner, sender;
18   socklen_t addr_len;
19     sockfd=socket(AF_INET,SOCK_DGRAM,0);
20
21   if(sockfd < 0){
22                   perror("\n Error in opening socket ... ");
23                   exit(1);
24   }
25
26   listner.sin_family = AF_INET;
27   listner.sin_addr.s_addr = htons(INADDR_ANY);
28   listner.sin_port = htons(MYPORT);
29
30   if(bind(sockfd, (struct sockaddr *)&listner, sizeof(listner)))
      {
31                   perror("\n Error in bind ... ");
32                   exit(2);
33   }
34
35   printf("listener: waiting to recvfrom...\n");
36   addr_len = sizeof sender;
37
38   if ((numBytes = recvfrom(sockfd, buf, MAXBUFLEN-1, 0, (struct
      sockaddr *)&sender, &addr_len)) == -1) {
39     perror("recvfrom");
40     exit(1);
41   }
42
43   printf("got packet from: %s\n", inet_ntoa(sender.sin_addr));
44   printf("Packet contains: \"  %s \" \n", buf);
45
46   close(sockfd);
47   return 0;
48 }
```

The program in Listing 3.15 is any UDP server program. This program can receive either a broadcast or a unicast message and prints the same. In line 18 it creates the socket and in line 28 it binds the socket to a local protocol

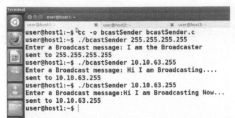

Figure 3.24: Broadcast sender terminal

Figure 3.25: Broadcast listener 1

Figure 3.26: Broadcast listener 2

Figure 3.27: Broadcast listener 1

address. The program now blocks at line 32 (*recvfrom()*) waiting for a *sendto()* call to be executed by any UDP client. In lines 36 to 37, it prints the IP address of the sender as well as the message received from it and then stops.

These programs must be executed on different machines of a single subnet to experience real broadcasting. The program execution is demonstrated in figures 3.24to 3.27. Each of these terminals belongs to a different computer in the same LAN, as can be seen in the command prompt (different host names) provided in the first line of the figures. To understand the execution of these programs, follow the steps below for execution with reference to the figures, where Figure 3.24 represents the broadcaster and Figures 3.25 to 3.27 refers to three broadcast listeners.

i) Copy the listener program code to more than one computer of a same LAN.

ii) Compile the program individually in respective computers, as shown in the first lines of Figures 3.25 to 3.27. Now run the programs in respective places as given in the second line.

iii) Similarly, copy and compile the sender program in host 1, and execute the sender, as shown in line 1 and line 2 of Figure 3.24, with a global broadcast address.

iv) Now enter a message in the host 1 terminal to broadcast. After broadcasting, the sender stops in line 5 of Figure 3.24.

v) The message is received by all listeners simultaneously and printed along with the sender's IP in line 4 and line 5 of Figures 3.25 to 3.27. After printing the message, the listener program stops in line 6.

vi) At this point, the sender again broadcasts a message to the subnet's broadcast address in line 5 of Figure 3.24. But, unfortunately, all the listeners are already stopped by this time; therefore the message is dropped at the switch.

vii) Again run the listeners in all the receiving hosts, as shown in line 6 of Figures 3.25 to 3.27.

viii) Run the sender again with a message to the subnet's IP, as in lines 8 and 9 of Figure 3.24.

ix) The broadcast message is again received and printed by all listeners as usual.

x) All programs are stopped.

Test/Viva Questions

a) Can a broadcast message be passed through an L2 switch?

b) Can a broadcast message be passed through an L3 switch?

c) What address is used as the receiver's address in broadcasting?

d) What is the difference between a global broadcast address and a subnet's broadcast address?

e) INADDR_BROADCAST represents which address?

f) Write the prototype for *setsockopt()*. Why is this function used in a broadcast sender?

g) Write and explain the parameters needed for *getsockopt()*.

h) Which operating system command can be used to get the subnet's broadcast address?

i) Justify that a broadcasting sender needs to be a UDP client.

j) How many UDP servers/clients are required in a broadcasting system?

k) Is it possible to use a TCP socket for broadcasting? Justify your answer.

Programming Assignments

1) Listing 3.14 needs broadcast IP in the command line. Modify this program to hardcode IP to INADDR_BROADCAST in the program.

2) Modify the code in Listing 3.14 to take an extra command line argument for *time* in seconds. The sender should broadcast the message repeatedly after each *time* interval.

Bibliography

[1] W. Richard Stevens, *Unix Network Programming*, Vol-1, PHI (Prentice Hall of India).

[2] Beej's Guide to Network Programming—Using Internet Sockets, http://beej.us/guide/bgnet, last accessed on 1 Aug 2013.

[3] M. J. Donahoo and K. L. Calvert, *TCP/IP Sockets in C*, Morgan Kaufmann, USA.

CHAPTER 4

INTRODUCTION TO NS2

Network Simulator version 2 (NS2) is an open source tool that was designed to help in network design and research. It provides substantial support for simulation of almost all existing protocols over wired and wireless (both local and satellite) networks in different configurations. The elegance of this tool lies in the support for a framework to study design and evaluate any new protocol. By framework we mean the following. A network simulation framework considerably simplifies protocol simulation for the following reason.

Any network protocol works by making use of the service provided by a set of lower layer protocols in a layered protocol stack. Also a protocol is invoked by higher layer protocols. These protocols in the network stack are already built into NS2. Without these supporting protocols, one would have to implement all these supporting protocols to perform simulation of a single protocol.

NS2 was created in 1989 as a variant of the REAL network simulator. It got support from DARPA in 1995, which accelerated its development. It has been developed as an academic project over last two decades and is currently maintained by Information Sciences Institute (ISI), University of Southern California. Currently, the tool has more then 300K lines of code, with a 400-page manual. It has a very large user base that is spread a round the globe and encompasses both industry users and academia. It has become a de facto standard in networking study and research because of its easy and free availability and flexibility of incorporating new designs according to user requirements. Since the protocols at different layers are already inbuilt in NS2, a protocol developer developing protocol at any layer can effortlessly simulate it, since all the supporting protocols have already been programmed into it.

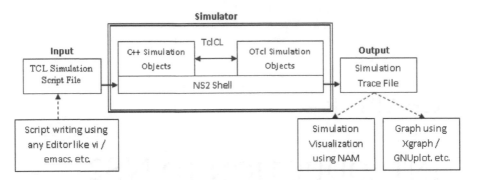

Figure 4.1: Basic structure of NS2

4.1   Simulator Structure

NS2 has been programmed using two object-oriented languages, C++ and
Object Oriented Tool Command Language (OTcl). OTcl programs are written
by the user and act as the front end, while C++ programs act as the back end
that executes the actual simulation. Figure 4.1 shows the basic structure of
NS2. As can be seen in Figure 4.1, actual protocols have been implemented
internally using C++, whereas OTcl provides a handle to the user to use these
protocols. The connection between OTcl and C++ has been made by an interface called TclCL (Tcl with Classes Library). It provides a layer of C++ glue
over OTcl.

NS2 provides a large number of built-in C++ objects that can be used
to set up a basic simulation using a Tcl simulation script. However, an advanced user intending to design a new protocol needs to develop his own
C++ objects and has to design an OTcl configuration interface to access newly
designed objects via a Tcl script. C++ code runs faster than OTcl code, but
changes to C++ code require more time and effort as compared to changes in
OTcl code because C++ code is compiled, while OTcl code is interpreted, i.e.,
a small change (say int i=0 is changed to i=1) in C++ code would compile
the complete package, where an interpreter can run only the changed line. In
practice a user needs to change the input values (configuration) frequently
while testing a simulation experiment. Therefore OTcl helps to perform a
large number of simulation experiments with different network configurations, for example, a different number of nodes and links can be created using OTcl.

Both C++ and OTcl programs consist of two different class hierarchies,
the linked hierarchy and the standalone hierarchy. The classes of the linked
hierarchy of both the domains are linked together using TclCL as shown in
Figure 4.2. In other words, for each class available in the C++ hierarchy, a
corresponding class also exists in OTcl, thereby setting up a one-one correspondence. The OTcl linked hierarchy is called the "*interpreted hierarchy*" and

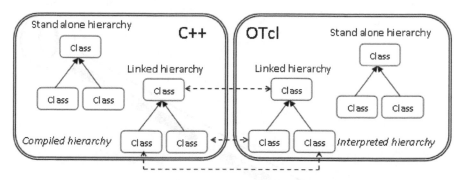

Figure 4.2: Interlinked C++ and OTcl domains

the counterpart C++ hierarchy is called the *"compiled hierarchy"* (see Figure 4.2). The second type of hierarchy includes classes from both the domains that are not linked together. These classes are neither a part of the interpreted hierarchy nor a part of the compiled hierarchy. Users may have to program using any one (both) language according to the specific network simulation problem at hand. However, the following is a guideline as to when to use OTcl and when to use C++.

- Use OTcl:
  - For network configuration and setup
  - To run simulation with existing NS2 modules
- Use C++:
  - To modify one or more existing modules or to create new modules to implement new protocols.

This text focuses on setting up and performing network simulation experiments using OTcl. The aim is largely restricted to understanding the performance and behavior of existing protocols in different network settings through simulation exercises.

4.2 Simulator Input and Output

Input to the simulator is a TCL script written as a text file and interpreted by the NS2 shell. The simulator normally produces two files (containing ASCII text) as output: a NAM trace file and an NS or packet trace file. We explain the purpose of these two files in the following.

The NAM trace file is used as input to another package called Network Animator (NAM) that animates the complete simulation trace in a graphical user interface. An example of NAM visualization is shown in Figure 4.3,

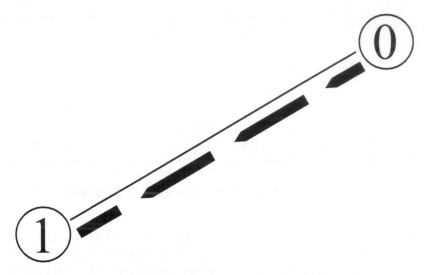

Figure 4.3: Example visualization produced by NAM

where two nodes are communicating via a point-to-point link. The advantage of using animation in visualization is twofold. The user can easily verify that the topology designed by him is actually what he wanted, and it increases the understandability of the complete process.

The second output file is called the NS or packet trace file. It can be used to study network performance parameters like delay, throughput, etc. The trace file usually contains multiple rows, and each row represents an event that has occurred during the simulation experiment. These rows are produced in a predefined format from which the required data can be extracted using any text processing tool (written using a script or some programming language). After data have been collected by using any tool, any of the graph drawing tools may be used to plot the graphs. Some tools that are popularly used to plot the collected data points are Xgraph, Gnuplot, MATLAB®, MS Excel, etc. An example graph is shown in Figure 4.4. Even while using these tools, it is still tedious to extract the data from the trace files, as no ready-made package is provided. While conducting your simulation experiments, you may use shell scripts, AWK scripts, Perl or C/C++ programs to extract the required information from a trace file.

4.3   NS2 Installation Steps

NS2 is an open source software package available freely (subject to open source licence agreements) on the web. It can run on Unix (or Linux), Windows, as well as Mac systems. It is easier to use the software in the Unix environment, as it was developed in this environment. Unless otherwise

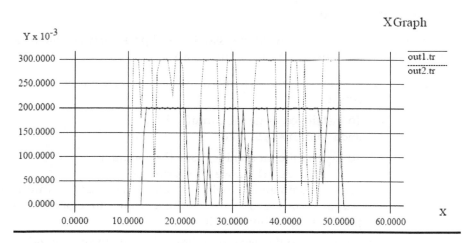

Figure 4.4: Example graph produced by Xgraph

specified, all subsequent discussions in this text are with reference to Unix/Linux systems.

All the components of the NS2 source are available as either one bundle called an all-in-one package or different components can be downloaded and installed separately. The current version of the software at the time of writing is 2.35 and might be upgraded by the time the book reaches the reader. However, the basic functionalities this book addresses are expected to remain more or less the same in future versions. The current all-in-one suite consists of the following modules.

- `NS release 2.35`

- `Tcl/Tk release 8.5.10`

- `OTcl release 1.14`

- `TclCL release 1.20`

There are also a few optional modules.

- `NAM release 1.15`

- `Zlib version 1.2.3`

- `Xgraph version 12`

Installation steps for different operating system distributions are provided below as examples. If the reader is interested in other platforms, they are directed to follow the instructions provided on NS's official website [1].

Download the package `ns-allinone-2.35.tar.gz` from the website (`http://sourceforge.net/projects/nsnam/files/allinone/ns-allinone-2.35/`) to the folder where you want to install.

4.3.1   Fedora Linux

Log in to a user where you want to install the NS2 package (say myuser) and
follow the steps to install NS2 in Fedora Linux.

Step I: 1) Log in to a user where you want to install the NS2 package
(say myuser)

2) To install gcc, issue the command
`sudo yum install gcc`

3) To expand the tar archive, issue the command
`tar -xvzf ns-allinone-2.35.tar.gz`
This command will create a new directory
'ns-allinone-2.35' under the current directory and expand the tar archive into it.

4) To change to the expanded directory, issue the command
`cd ns-allinone-2.35`

5) Open the file ns-2.35/linkstate/ls.h, move to line 137, and
change the line
`void eraseAll() { erase(baseMap::begin(),`
`baseMap::end()); }`
to
`void eraseAll() { this->erase(baseMap::begin(),`
`baseMap::end()); }`
Now save and exit from the file.

6) Now to install the package, issue the command
`./install`
Now you have to wait for a long time and observe a whole lot
of text. If everything goes right, then finally you should see the
installation finish-up text shown in Figure 4.5.

Step II: If the package is installed successfully, then open the file
~/.bash_profile (or .profile or .bashrc or .login) in any text editor
(i.e., vi, gedit, etc.) and add the path information as given in Figure
4.6 at the end of the file.

– Save and exit from the editor

– Activate the path information by issuing the following command:
`source ~/.bash_profile`

Step III: You can now test the installation by issuing the following command:

– change to the NS2 directory
`cd ns-2.35`

```
Ns-allinone package has been installed successfully.
Here are the installation places:
tcl8.5.10:       /home/myuser/ns-allinone-2.35/{bin,include,lib}
tk8.5.10:        /home/myuser/ns-allinone-2.35/{bin,include,lib}
otcl:            /home/myuser/ns-allinone-2.35/otcl-1.14
tclcl:           /home/myuser/ns-allinone-2.35/tclcl-1.20
ns:              /home/myuser/ns-allinone-2.35/ns-2.35/ns
nam:    /home/myuser/ns-allinone-2.35/nam-1.15/nam
xgraph: /home/myuser/ns-allinone-2.35/xgraph-12.1
gt-itm:    /home/myuser/ns-allinone-2.35/itm, edriver, sgb2alt,
sgb2ns, sgb2comns, sgb2hierns


---------------------------------------------------------


Please put /home/myuser/ns-allinone-2.35/bin:/home/myuser/ns-
allinone-2.35/tcl8.5.10/unix:/home/myuser/ns-allinone-
2.35/tk8.5.10/unix
into your PATH environment; so that you'll be able to run
itm/tclsh/wish/xgraph.

IMPORTANT NOTICES:

(1) You MUST put /home/myuser/ns-allinone-2.35/otcl-1.14,
/home/myuser/ns-allinone-2.35/lib,
    into your LD_LIBRARY_PATH environment variable.
    If it complains about X libraries, add path to your X libraries
    into LD_LIBRARY_PATH.
    If you are using csh, you can set it like:
            setenv LD_LIBRARY_PATH <paths>
    If you are using sh, you can set it like:
            export LD_LIBRARY_PATH=<paths>

(2) You MUST put /home/myuser/ns-allinone-2.35/tcl8.5.10/library
into your TCL_LIBRARY environmental
    variable. Otherwise ns/nam will complain during startup.

(3) [OPTIONAL] To save disk space, you can now delete directories
tcl8.5.10
    and tk8.5.10. They are now installed under /home/myuser/ns-
allinone-2.35/{bin,include,lib}

After these steps, you can now run the ns validation suite with
cd ns-2.35; ./validate

For trouble shooting, please first read ns problems page
http://www.isi.edu/nsnam/ns/ns-problems.html. Also search the ns
mailing list archive for related posts.
```

Figure 4.5: NS2 installation finish-up text

    – validate the package
     `./validate`
     Validation takes a rather long time. If it runs without any problem, then you probably have a good installation.

Or execute sample programs by issuing the following commands

```
PATH=$PATH:/home/myuser/ns-allinone-2.35/bin:
/home/myuser/ns-allinone-2.35/tcl8.5.10/unix:
/home/myuser /ns-allinone-2.35/tk8.5.10/unix

LD_LIBRARY_PATH=$LD_LIBRARY_PATH:/home/myuser/ns-
allinone-2.35/otcl-1.14:/home/myuser/ns-allinone-2.35/lib

TCL_LIBRARY=$TCL_LIBRARY:/home/myuser/ns-allinone-
2.35/tcl8.5.10/library

export PATH
export LD_LIBRARY_PATH
export TCL_LIBRARY
```

Figure 4.6: Path information

– change to the example directory
  cd /home/myuser/ns-allinone-2.35/ns-2.35/tcl/ex
– run the simulation script *simple.tcl*
  ns simple.tcl

4.3.2 Ubuntu Linux

Log in to a user where you want to install the NS2 package (say myuser) and follow the steps to install NS2 in Ubuntu Linux.

Step I: 1) First update the online software repositories
             sudo apt-get update

2) Install the c++ compiler and other required packages as follows:
   sudo apt-get install g++ build-essential autoconf
   automake libxmu-dev

3) To expand the tar archive, issue the command
   tar -xvzf ns-allinone-2.35.tar.gz
   This command will create a new directory 'ns-allinone-2.35' under the current directory and expand the tar archive into it.

4) To change into the expanded directory, issue the command
   cd ns-allinone-2.35

5) Open the file ns-2.35/linkstate/ls.h, move to line 137, and change the line
   void eraseAll() { erase(baseMap::begin(),
   baseMap::end()); }
   to
   void eraseAll() { this->erase(baseMap::begin(),

```
baseMap::end()); }
```
Now save and exit from the file.

6) Now to install the package, issue the command
   ```
   ./install
   ```
   Now you have to wait for a long time and observe a whole lot of text. If everything goes right, finally you should see the installation finish-up text shown in Figure 4.5.

Step II: If the package is installed successfully, then open the file ~/.bashrc in any text editor (i.e., vi, gedit, etc.) and add the path information as given in Figure 4.6 at the end of the file.

- Save and exit from the editor
- Activate the path information by issuing the following command:
  ```
  source ~/.bashrc
  ```

Step III: You can now test the installation by issuing the following command:

- change to the NS2 directory
  ```
  cd ns-2.35
  ```
- validate the package
  ```
  ./validate
  ```
  Validation takes a rather long time. If it runs without any problem, then you probably have a good installation.

Or execute sample programs by issuing the following commands

- change to the example directory
  ```
  cd /home/myuser/ns-allinone-2.35/ns-2.35/tcl/ex
  ```
- run the simulation script *simple.tcl*
  ```
  ns simple.tcl
  ```

4.3.3 Windows 7

Installing NS2 in a Windows operating system is a bit tricky and lengthy. We provide here a step by step method of installation. You can also find a nice guide to install NS2 in Windows at `http://www.net.c.dendai.ac.jp/ns2/`. To install the NS2 package in a Windows operating system, you need to first install the *cygwin* package freely available from `https://cygwin.com/install.html`. Cygwin provides a Unix-like environment and command-line interface for Microsoft Windows.

Step I: Download `setup-x86.exe` for 32-bit versions of Windows from the above website and run it. It asks for some options while installing. Some of the major options are as follows.

    i) Choose A Download source: Install from the Internet

    ii) Select Root Installation Directory: c:\cygwin

   iii) Select Local Package Directory: c:\cygwin

   iv) Select Your Internet Connection: choose your connection

    v) Choose a Download Site: choose any from the list provided

   vi) Select Packages: In this step select some extra packages needed for NS2 and NAM, as given in Table 4.1. Column 1 represents the required packages and column 2 provides the category under which these packages are available. However, if some packages are not found in the said category, then use the search box provided in the top left corner of the "Select Packages" window.

Once all the packages are selected, click next to install. It will take some time to finish the installation. Finally, choose to have a desktop icon of cygwin.

Table 4.1: Extra packages selection in Cygwin

Package name	Category
gcc-core	Devel
gcc-g++	Devel
make	Devel
patch	Devel
w32api-header	Devel
w32api-runtime	Devel
perl	Interpreter
tcl	Interpreter
tcl-devel	Interpreter
tcl-tk	Interpreter
tcl-tk-devel	Interpreter
vim	Editor
gedit	Editor
xorg-server	X11
xinit	X11
libx11-devel	Lib
libxmu-devel	Lib

Step II: Copy ns-allinone-2.35.tar.gz to the c:\cygwin\home\myuser folder. Note that myuser must be replaced with your user in cygwin, which will be found under the home directory.

   1) To expand the tar archive, issue the command

```
tar -xvzf ns-allinone-2.35.tar.gz
```
This command will create a new directory 'ns-allinone-2.35' under the current directory and expand the tar archive into it.

2) To change into the expanded directory, issue the command
`cd ns-allinone-2.35`

3) Remove the `tcl` and `tk` folders from here as they have been installed with the cygwin installation
`rm -rf tcl8.5.10`
`rm -rf tk8.5.10`

4) Open the file /usr/include/tcl8.5/generic/tclPort.h, go to line 23, and change the line
`# include "tclUnixPort.h"`
to
`# include "../unix/tclUnixPort.h"`

5) Change the name of libraries by making symbolic links by issuing the following commands:
`ln -s /usr/lib/libtcl.dll.a /usr/lib/libtcl.a`
`ln -s /usr/lib/libtk.dll.a /usr/lib/libtk.a`

6) You need to modify/patch the `install` file provided in the `ns-allinone-2.35` folder and some other configuration files. To do that either create a patch file (pfile.txt) by entering the content provided in the appendix at the end of this chapter into a text file or download the file from the website `http://www.net.c.dendai.ac.jp/ns2/`. Once you have created/downloaded the patch file, copy it to the folder containing the folder `ns-allinone-2.35` (`c:\cygwin\home\myuser` in this case). Issue the following command to patch the files.
if you are inside the `ns-allinone-2.35` then change to parent
`cd../`
`patch -p0 < pfile.txt`

7) Now you are ready to install the package and issue the command
change the directory as follows.
`cd ns-allinone-2.35`
`./install`
Now you have to wait for a long time and observe a whole lot of text. If everything goes right, then you should see the installation finish-up text shown in Figure 4.5.

Step III: If the package is installed successfully, then open the file ~/`.bash_profile` in any text editor and add the path information at the end of the file as follows.
`PATH="${HOME}/ns-allinone-2.35/bin:${PATH}"`

– Save and exit from the editor

– Activate the path information by issuing the following command:

```
source ~/.bash_profile
```
Again, to run Windows programs like NAM, gedit, etc., you
need to issue the following commands.
```
startxwin &
export DISPLAY=:0.0
```
If you want to make it permanent, then add these two lines at
the end of ~/.bash_profile

Step IV: You can now test the installation by executing a sample program
by issuing the following command.
```
ns/home/myuser/ns-allinone-2.35/ns-2.35/tcl/
ex/simple.tcl
```

4.4   NS2 Directories and Files

Let us take a tour of the NS2 software just installed. The directory structure
is shown schematically in Figure 4.7. You can view the subdirectories and
files contained in them using the cd and ls commands. In this figure only
the important folders are shown.

- Level 2
  ns-2.35: This contains all the NS2 simulation modules. That is, all the

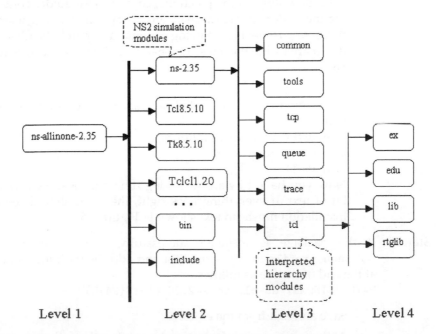

Figure 4.7: NS2 directory hierarchy

C++ modules as well as tcl modules are present here. Hereinafter ~ns refers to this folder, i.e., /home/myuser/ns-allinone-2.35/ns-2.35/
tcl8.5.10: contains the tcl package
tk8.5.10: contains the tk package
nam1.15: contains the NAM package
tclcl1.20: contains classes in TclCL
xgraph12.2: contains the xGraph package

. . .

bin: contains the executables of various packages like ns, nam, and xGraph, etc.
include: Some useful headers are provided in this folder for Tcl and Tk.

- Level 3
common: The folders of this level normally contain common packet forwarding modules such as the simulator and scheduler, etc.
queue: contains queuing modules
tcp: contains the Transmission Control Protocol
trace: contain modules for tracing
tcl: contains interpreted hierarchy

. . .

- Level 4
edu: contains some educational ns scripts to demonstrate networking concepts like "hidden and exposed terminal problems of Wireless LAN," etc.
ex: contains many example ns scripts, which could be executed and observed
lib: contains the OTcl code

. . .

4.5   Network Animator (NAM)

The NAM package comes with the software package called the ns-allinone bundle (tarball). This is a visualization tool that animates a network simulation. It has been developed using Tcl/TK and is used to view different network elements such as the nodes, links, and the network events such as packet transmission, packet drop, node movement, etc.

The input to this tool is the NAM trace file generated as an output of an NS simulation experiment. This trace file is in fact a simple text file containing different fields like topology information and packet traces. The NAM tool creates visualization based on the information contained in the NAM trace file. It also provides a user interface that provides control over different aspects of animation, such as topology, simulation speed, pause animation, etc.

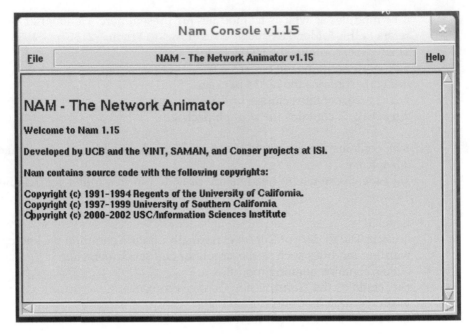

Figure 4.8: NAM console

To start NAM, issue the command *nam* from the command line. This will start the NAM console window shown in Figure 4.8. Multiple animations may run under the same nam instance. This window has two menu items. The corresponding submenu items are as follows.

1. **File** menu: This has the following submenu.
   New Nam editor: This should help to create an NS topology using the nam editor.
   Open: This option is used to open existing trace files.
   WinList: This command pops up a window containing the names of all currently opened tracefiles.
   Quit: This helps to quit from NAM.

2. **Help** menu: This has the following submenu items.
   Help: Contains a single page help screen.
   About: Shows version and copyright information.

A trace file can be opened either from the *open* submenu in the **file** menu of the NAM console window, or by using the trace file name as the command line argument as

$ nam <nam_trace>.nam

The animation window comes up, as shown in Figure 4.9. This is the main activity window having the following components.

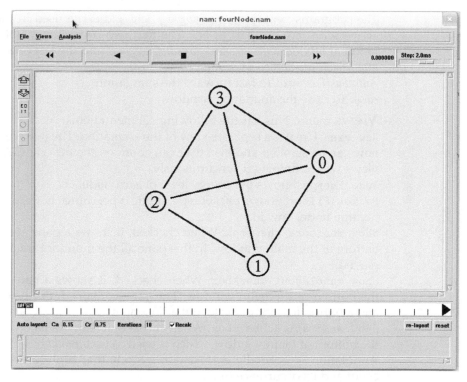

Figure 4.9: NAM window

- Menu bar

    - **File** menu: This has the following submenu items.
      Save layout: To save the current network layout to a file.
      Print: print the current network layout.
      Record Animation: records the animation.
      It is used to save frames of animation play. When clicked, each an-
      imation frame is saved in a file named *nam%d.xwd*, and %d is re-
      placed with the frame number. All of these files containing frames
      may be combined to form an animated GIF or MPEG using some
      third-party post-processing tools such as xwdtoppm, ppmtogif,
      and gifmerge.
      We now provide an example shell script to combine frames:

```
                Listing 4.1: Animated gif conversion
1  for i in *.xwd; do
2    xwdtoppm <$i | ppmtogif -interlace -transparent
3    '#e5e5e5' > `basename \$i .xwd`.gif;
4  done
5  gifmerge -10 -2 -229,229,229 *.gif > movie.gif
```

The programs xwdtoppm, ppmtogif, and gifmerge used in the above script are not part of NS. The first two programs may be found at *http://download.sourceforge.net/netpbm/* and the last one at *http://www.the-labs.com/GIFMerge/*.

Auto FastForward: To fast forward the simulation.

close: To close the animation window.

- **Views** menu: This has the following submenu items

  New view: Creates a separate view of the animation. The newview now can be scrolled and also user can zoom on the newview. All views will be animated synchronously.

  Node Energy: Shows the energy level of each node.

  Packet Filter: Shows packet type, traffic type, source node, destination node, flow id.

  Show monitors checkbox: When checked, it shows a pane at the bottom of the main window. In this pane all the monitors are displayed.

  Show autolayout checkbox: When checked, it shows a pane at the bottom of the main window. The pane contains input boxes and a button for automatic layout adjustments.

  Show annotation checkbox: When checked, it shows a listbox at the bottom of main window, which is used to list *annotations*.

  Annotation is normally a *(time, string)* pair that represents an event and its occurrence time.

- **Control Bar**: This is present below the menu bar of the NAM window. The different buttons available on the control bar (left to right) are described below.

  - (<<) — *Rewind*. Click on this button to move the animation time backward.

  - (<) — *Backward play*. Click on this button to play the animation backward.

  - (□) — *Stop*. Click on this button to pause the animation.

  - (>) — *Forward play*. Click on this button to play the animation forward.

  - (>>) — *Fast forward*. Click on this button to move the animation time forward.

- Time label — This displays the total time of animation. This value is taken from the simulation time as found in the trace file.

- Rate Slider — This button adjusts the animation granularity by specifying the screen update rate. When the button slides to the right, the animation rate increases and vice versa.

- *Tool bar* (left side of animation window) contains two buttons for zoom in (up arrow) and zoom out (down-arrow), respectively.

- *Scroll bars* (right and bottom of animation window) — These are used to slide the main animation view.

- *Time Slider* (below animation area) — This looks like a ruler with a scale. It also contains a tag TIME that can slide along the scaled ruler. Normally the tag slides automatically along with the animation play. However, the user can drag the tag manually to play either forward or backward in time. It has the same effect as rewind or fast forward, depending on the drag direction.

Besides these, the color, orientation of the node's physical position, links, and some other functionalities can also be manipulated through Tcl scripts in the simulation program. These capabilities are discussed in Chapters 5 and 6.

4.6   NS2 Program Structure

The NS2 program is basically a Tcl/OTcl script written using any text editor such as vi or emacs. The normal program structure is as follows:

- Program description as comment (optional)

- Create a Simulator Object

- Create NS2 trace file

- Create NAM trace file

- Create topology (define nodes and links with permissible characteristics like simplex/duplex, bandwidth, etc.)

- Create a transport agent in pair (both source and sink), attach them to corresponding nodes, and connect them

- Create application/traffic agent and attach it to the required nodes

- Schedule different events (when to start and stop, etc.)

- Define a finish procedure

- Start the simulation

The normal finish procedure consists of the following steps:

- Declare simulator object and trace files as global

- Flush the traces

- Close trace files (both NS and NAM)

- Execute NAM with NAM trace file as input (if any)

Example 4.1:
Consider an example topology that contains two nodes (n0 and n1) and a duplex link between them, as shown in Figure 4.3. It uses TCP as the transport layer protocol (called transport agent) and an ftp traffic is sent from node n0 to node n1.

Listing 4.2: First NS2 program myFirst.tcl

```
1  # Create a Simulator
2  set ns [new Simulator]
3
4  # Create a trace file
5  set mytrace [open out.tr w]
6  $ns trace-all $mytrace
7
8  # Create a NAM trace file
9  set myNAM [open out.nam w]
10 $ns namtrace-all $myNAM
11
12 # Create Nodes
13 set n0 [$ns node]
14 set n1 [$ns node]
15
16 # Connect Nodes with Links
17 $ns duplex-link $n0 $n1 100Mb 5ms DropTail
18
19 # Create a TCP agent
20 set tcp [new Agent/TCP]
21 $ns attach-agent $n0 $tcp
22 set sink [new Agent/TCPSink]
23 $ns attach-agent $n1 $sink
24 $ns connect $tcp $sink
25
26 # Create an FTP session
27 set ftp [new Application/FTP]
28 $ftp attach-agent $tcp
29
30 #Schedule events
31 $ns at 1.0 "$ftp start"
32 $ns at 30.0 "$ftp stop"
33 $ns at 30.1 "finish"
34
35 #Define a procedure finish
36 proc finish { } {
37   global ns mytrace myNAM
38   $ns flush-trace
39   close $mytrace
40   close $myNAM
41   exec nam out.nam &
42   exit 0
43 }
44
```

```
45  # Start the simulation
46  $ns run
```

The complete program is given in Listing 4.2. Write the program in a text file and save it as myFirst.tcl. Now execute the following command.
$ ns myFirst.tcl
When the simulation finishes, the NAM will automatically start up. Now press the run button of NAM to experience the visualization. A more detailed explanation of the Tcl simulation script is provided in Chapter 6.

In the following, we give the learning steps to be followed to perform a network simulation experiment.

Basic Simulation Steps

i) Learn Tcl/OTcl scripting, which is required to write NS2 programs

ii) Learn AWK scripting, which is required to extract data from an output trace file.

iii) Learn Gnuplot to plot graphs.

iv) Write NS2 programs using Tcl and execute. The program execution will produce event trace files.

v) Extract required information from trace files using AWK and then the values extracted may be plotted using Gnuplot.

Note: Steps iv and v above are discussed in more detail in Chapter 6.

4.7   Summary

We have discussed certain basic concepts of the NS2 tool briefly in this chapter, including the overall structure of the package. We have also discussed the interlink between the C++ and OTcl domains. The installation steps have been specified (restricted to Linux only) to help novices get the software ready. Post-processing tools such as NAM for animation, AWK to extract useful information from the output trace file, and Gnuplot to draw a graph to visualize and/or evaluate the performance of a simulation have been introduced. Finally, we discussed the general program structure and a basic example for a quick introduction to NS2.

Test/Viva Questions

a) What are the constituent parts of NS2?

b) What purpose can be achieved by Tcl/OTcl in NS2?

c) For what purpose C++ is used in NS2?

d) In the NS2 software, which folder contains the simulation examples?

e) What is the use of AWK and Gnuplot in NS2-based network simulation?

f) Explain the overall structure of NS2 using a block diagram.

g) Explain the reason for the use of two different languages in NS2 for conducting simulation experiments: OTcl and C++.

Tasks

1) Download and install NS2 in a Linux system and configure the paths. Download some tutorials.

2) Explore the directories inside the main NS2 directory.

3) Execute the program in Listing 4.2 and observe the output.

Bibliography

[1] NS2 official website (software, documentation): http://www.isi.edu/nsnam/ns/

[2] NS2 Wiki: http://nsnam.isi.edu/nsnam/index.php/Main_ Page

[3] Marc Greies, Tutorial for the network simulator ns: http://www.isi.edu/nsnam/ns/tutorial/

[4] Altman and Jimenez, NS for beginners: http://www-sop.inria.fr/members/Eitan.Altman/COURS-NS/n3.pdf

[5] Jae Chung and Mark Claypool, NS by Example, http://nile.wpi.edu/NS/Advanced Reading

[6] Teerawat Issariyakul Ekram Hossain, *Introduction to Network Simulator NS2*, Springer, 2008 (Chapters 1–3).

APPENDIX

Listing 4.3: PATCH file (pfile.txt)

```
 1  *** ns-allinone-2.35/install.orig 2015-07-21 02:04:02.796613200
       +0900
 2  --- ns-allinone-2.35/install    2015-07-21 03:40:45.544292000
       +0900
 3  **************
 4  *** 204,210 ****
 5    fi
 6
 7    # A Cygwin install requires these packages:
 8  ! packages_base="gcc4 gcc4-g++ gawk tar gzip make patch perl
       w32api"
 9    packages_xorg="xorg-server xinit libX11-devel libXmu-devel"
10
11    # Check if we are using Cygwin, and if so, if it is a good
       install
12  --- 204,210 ----
13    fi
14
15    # A Cygwin install requires these packages:
16  ! packages_base="gcc-core gcc-g++ gawk tar gzip make patch perl
       w32api-headers w32api-runtime tcl tcl-devel tcl-tk tcl-tk-
       devel"
17    packages_xorg="xorg-server xinit libX11-devel libXmu-devel"
18
19    # Check if we are using Cygwin, and if so, if it is a good
       install
20  **************
21  *** 267,272 ****
22  --- 267,273 ----
23    echo "* Build XGraph-$XGRAPHVER"
24    echo "=========================="
25
26  + if [ -d ./xgraph-$XGRAPHVER ] ; then
27    cd ./xgraph-$XGRAPHVER
28    ./configure --prefix=../
29    if [ "${test_cygwin}" = "true" ]; then
30  **************
31  *** 280,285 ****
32  --- 281,287 ----
33    fi
34
35    cd ../
36  + fi
37
38    # Compile and install cweb and sgblib
39
40  **************
41  *** 399,404 ****
42  --- 401,407 ----
43    echo "* Build tcl$TCLVER"
44    echo "===================="
45
```

```
46 + if [ -d ./tcl$TCLVER ] ; then
47   cd ./tcl$TCLVER/unix
48   if [ -f Makefile ] ; then
49     make distclean
50 ***************
51 *** 422,427 ****
52 --- 425,431 ----
53   fi
54
55   cd ../../
56 + fi
57
58   # compile and install tk
59
60 ***************
61 *** 429,434 ****
62 --- 433,439 ----
63   echo "* Build Tk$TKVER"
64   echo "====================="
65
66 + if [ -d ./tk$TKVER ] ; then
67   cd ./tk$TKVER/unix
68   if [ -f Makefile ] ; then
69     make distclean
70 ***************
71 *** 451,456 ****
72 --- 456,462 ----
73   fi
74
75   cd ../../
76 + fi
77
78   #
79   # Since our configures search for tclsh in $PATH, the
         following
80 *** ns-allinone-2.35/otcl-1.14/configure.orig 2015-07-21
       03:12:47.543719900 +0900
81 --- ns-allinone-2.35/otcl-1.14/configure   2015-07-21
       03:51:28.973713000 +0900
82 ***************
83 *** 4791,4797 ****
84   if test "${with_tcl+set}" = set; then
85     withval=$with_tcl; d=$withval
86   else
87 !   d=""
88   fi
89
90
91 --- 4791,4797 ----
92   if test "${with_tcl+set}" = set; then
93     withval=$with_tcl; d=$withval
94   else
95 !   d="/usr"
96   fi
97
98
```

```
 99  ***************
100  *** 4896,4903 ****
101  --- 4896,4906 ----
102    TCL_H_PLACES_D="$d/generic \
103        $d/unix \
104        $d/include/tcl$TCL_HI_VERS \
105  +     $d/include/tcl$TCL_HI_VERS/generic \
106        $d/include/tcl$TCL_VERS \
107  +     $d/include/tcl$TCL_VERS/generic \
108        $d/include/tcl$TCL_ALT_VERS \
109  +     $d/include/tcl$TCL_ALT_VERS/generic \
110        $d/include \
111        /usr/local/include \
112        "
113  ***************
114  *** 5406,5412 ****
115    if test "${with_tk+set}" = set; then
116      withval=$with_tk; d=$withval
117    else
118  !   d=""
119    fi
120
121
122  --- 5409,5415 ----
123    if test "${with_tk+set}" = set; then
124      withval=$with_tk; d=$withval
125    else
126  !   d="/usr"
127    fi
128
129
130  ***************
131  *** 5503,5510 ****
132  --- 5506,5516 ----
133                    $d/../include/tk$TK_HI_VERS \
134                    $d/../include/tk$TK_OLD_VERS \
135        $d/include/tk$TK_VERS \
136  +     $d/include/tk$TK_VERS/generic \
137        $d/include/tk$TK_HI_VERS \
138  +     $d/include/tk$TK_HI_VERS/generic \
139        $d/include/tk$TK_OLD_VERS \
140  +     $d/include/tk$TK_OLD_VERS/generic \
141                    $d/include"
142    TK_H_PLACES=" \
143        ../include \
144  *** ns-allinone-2.35/otcl-1.14/conf/configure.in.tcl.orig
145        2015-07-21 03:10:53.753678700 +0900
146  --- ns-allinone-2.35/otcl-1.14/conf/configure.in.tcl   2015-07-21
147        03:11:49.462671500 +0900
146  ***************
147  *** 1,7 ****
148    dnl autoconf rules to find tcl
149    dnl $Header: /cvsroot/otcl-tclcl/conf/configure.in.tcl,v 1.49
150      2009/12/30 19:36:09 tom_henderson Exp $ (LBL)
151  ! AC_ARG_WITH(tcl,  --with-tcl=path specify a pathname for tcl,
```

```
        d=$withval, d="")
152
153   dnl cant easily escape brackets in M4/autoconf-- must use
        quadigraphs below
154   AC_ARG_WITH(tcl-ver, --with-tcl-ver=path specify the version
        of tcl/tk, TCL_VERS=$withval, TCL_VERS=`echo "puts @<:@info
        patchlevel@:>@" | tclsh`)
155 --- 1,7 ----
156   dnl autoconf rules to find tcl
157   dnl $Header: /cvsroot/otcl-tclcl/conf/configure.in.tcl,v 1.49
        2009/12/30 19:36:09 tom_henderson Exp $ (LBL)
158
159 ! AC_ARG_WITH(tcl, --with-tcl=path specify a pathname for tcl,
        d=$withval, d="/usr")
160
161   dnl cant easily escape brackets in M4/autoconf-- must use
        quadigraphs below
162   AC_ARG_WITH(tcl-ver, --with-tcl-ver=path specify the version
        of tcl/tk, TCL_VERS=$withval, TCL_VERS=`echo "puts @<:@info
        patchlevel@:>@" | tclsh`)
163 ***************
164 *** 102,109 ****
165 --- 102,112 ----
166   TCL_H_PLACES_D="$d/generic \
167       $d/unix \
168       $d/include/tcl$TCL_HI_VERS \
169 +     $d/include/tcl$TCL_HI_VERS/generic \
170       $d/include/tcl$TCL_VERS \
171 +     $d/include/tcl$TCL_VERS/generic \
172       $d/include/tcl$TCL_ALT_VERS \
173 +     $d/include/tcl$TCL_ALT_VERS/generic \
174       $d/include \
175       /usr/local/include \
176       "
177 *** ns-allinone-2.35/otcl-1.14/conf/configure.in.tk.orig
        2015-07-21 03:49:25.300319900 +0900
178 --- ns-allinone-2.35/otcl-1.14/conf/configure.in.tk 2015-07-21
        03:50:01.208701600 +0900
179 ***************
180 *** 2,8 ****
181   dnl $Header: /cvsroot/otcl-tclcl/conf/configure.in.tk,v 1.34
        2009/12/30 19:36:09 tom_henderson Exp $ (LBL)
182
183
184 ! AC_ARG_WITH(tk, --with-tk=path  specify a pathname for tk, d=
        $withval, d="")
185
186   dnl library version defaults to 8.0
187   AC_ARG_WITH(tk-ver, --with-tk-ver=path specify the version of
        tcl/tk, TK_VERS=$withval, TK_VERS=$TCL_VERS)
188 --- 2,8 ----
189   dnl $Header: /cvsroot/otcl-tclcl/conf/configure.in.tk,v 1.34
        2009/12/30 19:36:09 tom_henderson Exp $ (LBL)
190
191
192 ! AC_ARG_WITH(tk, --with-tk=path  specify a pathname for tk, d=
```

```
193           $withval, d="/usr")
194     dnl library version defaults to 8.0
195     AC_ARG_WITH(tk-ver, --with-tk-ver=path specify the version of
        tcl/tk, TK_VERS=$withval, TK_VERS=$TCL_VERS)
196   ***************
197   *** 94,101 ****
198   --- 94,104 ----
199                   $d/../include/tk$TK_HI_VERS \
200                   $d/../include/tk$TK_OLD_VERS \
201       $d/include/tk$TK_VERS \
202   +   $d/include/tk$TK_VERS/generic \
203       $d/include/tk$TK_HI_VERS \
204   +   $d/include/tk$TK_HI_VERS/generic \
205       $d/include/tk$TK_OLD_VERS \
206   +   $d/include/tk$TK_OLD_VERS/generic \
207                   $d/include"
208     TK_H_PLACES=" \
209         ../include \
210   *** ns-allinone-2.35/tclcl-1.20/configure.orig   2015-07-21
        03:14:14.446092200 +0900
211   --- ns-allinone-2.35/tclcl-1.20/configure 2015-07-21
        03:53:02.443585700 +0900
212   ***************
213   *** 5914,5920 ****
214     if test "${with_tcl+set}" = set; then
215       withval=$with_tcl; d=$withval
216     else
217   !   d=""
218     fi
219
220
221   --- 5914,5920 ----
222     if test "${with_tcl+set}" = set; then
223       withval=$with_tcl; d=$withval
224     else
225   !   d="/usr"
226     fi
227
228
229   ***************
230   *** 6019,6026 ****
231   --- 6019,6029 ----
232     TCL_H_PLACES_D="$d/generic \
233         $d/unix \
234         $d/include/tcl$TCL_HI_VERS \
235   +     $d/include/tcl$TCL_HI_VERS/generic \
236         $d/include/tcl$TCL_VERS \
237   +     $d/include/tcl$TCL_VERS/generic \
238         $d/include/tcl$TCL_ALT_VERS \
239   +     $d/include/tcl$TCL_ALT_VERS/generic \
240         $d/include \
241         /usr/local/include \
242         "
243   ***************
244   *** 6529,6535 ****
```

```
245   if test "${with_tk+set}" = set; then
246     withval=$with_tk; d=$withval
247   else
248 !   d=""
249   fi
250
251
252 --- 6532,6538 ----
253   if test "${with_tk+set}" = set; then
254     withval=$with_tk; d=$withval
255   else
256 !   d="/usr"
257   fi
258
259
260 ***************
261 *** 6626,6633 ****
262 --- 6629,6639 ----
263                     $d/../include/tk$TK_HI_VERS \
264                     $d/../include/tk$TK_OLD_VERS \
265       $d/include/tk$TK_VERS \
266 +     $d/include/tk$TK_VERS/generic \
267       $d/include/tk$TK_HI_VERS \
268 +     $d/include/tk$TK_HI_VERS/generic \
269       $d/include/tk$TK_OLD_VERS \
270 +     $d/include/tk$TK_OLD_VERS/generic \
271                     $d/include"
272   TK_H_PLACES=" \
273       ../include \
274 *** ns-allinone-2.35/tclcl-1.20/conf/configure.in.tcl.orig
        2015-07-21 03:09:30.267029700 +0900
275 --- ns-allinone-2.35/tclcl-1.20/conf/configure.in.tcl 2015-07-21
        03:10:31.902085400 +0900
276 ***************
277 *** 1,7 ****
278   dnl autoconf rules to find tcl
279   dnl $Header: /cvsroot/otcl-tclcl/conf/configure.in.tcl,v 1.49
        2009/12/30 19:36:09 tom_henderson Exp $ (LBL)
280
281 ! AC_ARG_WITH(tcl, --with-tcl=path specify a pathname for tcl,
        d=$withval, d="")
282
283   dnl cant easily escape brackets in M4/autoconf-- must use
        quadigraphs below
284   AC_ARG_WITH(tcl-ver, --with-tcl-ver=path specify the version
        of tcl/tk, TCL_VERS=$withval, TCL_VERS=`echo "puts @<:@info
        patchlevel@:>@" | tclsh`)
285 --- 1,7 ----
286   dnl autoconf rules to find tcl
287   dnl $Header: /cvsroot/otcl-tclcl/conf/configure.in.tcl,v 1.49
        2009/12/30 19:36:09 tom_henderson Exp $ (LBL)
288
289 ! AC_ARG_WITH(tcl, --with-tcl=path specify a pathname for tcl,
        d=$withval, d="/usr")
290
291   dnl cant easily escape brackets in M4/autoconf-- must use
```

```
        quadigraphs below
292  AC_ARG_WITH(tcl-ver, --with-tcl-ver=path specify the version
        of tcl/tk, TCL_VERS=$withval, TCL_VERS=`echo "puts @<:@info
        patchlevel@:>@" | tclsh`)
293 ***************
294 *** 102,109 ****
295 --- 102,112 ----
296  TCL_H_PLACES_D="$d/generic \
297      $d/unix \
298      $d/include/tcl$TCL_HI_VERS \
299 +    $d/include/tcl$TCL_HI_VERS/generic \
300      $d/include/tcl$TCL_VERS \
301 +    $d/include/tcl$TCL_VERS/generic \
302      $d/include/tcl$TCL_ALT_VERS \
303 +    $d/include/tcl$TCL_ALT_VERS/generic \
304      $d/include \
305      /usr/local/include \
306      "
307 *** ns-allinone-2.35/tclcl-1.20/conf/configure.in.tk.orig
        2015-07-21 03:51:45.184152700 +0900
308 --- ns-allinone-2.35/tclcl-1.20/conf/configure.in.tk   2015-07-21
        03:52:20.950493900 +0900
309 **************
310 *** 2,8 ****
311  dnl $Header: /cvsroot/otcl-tclcl/conf/configure.in.tk,v 1.34
        2009/12/30 19:36:09 tom_henderson Exp $ (LBL)
312
313
314 ! AC_ARG_WITH(tk, --with-tk=path  specify a pathname for tk, d=
        $withval, d="")
315
316  dnl library version defaults to 8.0
317  AC_ARG_WITH(tk-ver, --with-tk-ver=path specify the version of
        tcl/tk, TK_VERS=$withval, TK_VERS=$TCL_VERS)
318 --- 2,8 ----
319  dnl $Header: /cvsroot/otcl-tclcl/conf/configure.in.tk,v 1.34
        2009/12/30 19:36:09 tom_henderson Exp $ (LBL)
320
321
322 ! AC_ARG_WITH(tk, --with-tk=path  specify a pathname for tk, d=
        $withval, d="/usr")
323
324  dnl library version defaults to 8.0
325  AC_ARG_WITH(tk-ver, --with-tk-ver=path specify the version of
        tcl/tk, TK_VERS=$withval, TK_VERS=$TCL_VERS)
326 **************
327 *** 94,101 ****
328 --- 94,104 ----
329              $d/../include/tk$TK_HI_VERS \
330              $d/../include/tk$TK_OLD_VERS \
331      $d/include/tk$TK_VERS \
332 +    $d/include/tk$TK_VERS/generic \
333      $d/include/tk$TK_HI_VERS \
334 +    $d/include/tk$TK_HI_VERS/generic \
335      $d/include/tk$TK_OLD_VERS \
336 +    $d/include/tk$TK_OLD_VERS/generic \
```

```
337                       $d/include"
338    TK_H_PLACES=" \
339          ../include \
340  *** ns-allinone-2.35/ns-2.35/configure.orig 2015-07-21
           03:17:29.999773500 +0900
341  --- ns-allinone-2.35/ns-2.35/configure   2015-07-21
           03:48:58.323859700 +0900
342  ***************
343  *** 5441,5447 ****
344    if test "${with_tcl+set}" = set; then :
345      withval=$with_tcl; d=$withval
346    else
347  !    d=""
348    fi
349
350
351  --- 5441,5447 ----
352    if test "${with_tcl+set}" = set; then :
353      withval=$with_tcl; d=$withval
354    else
355  !    d="/usr"
356    fi
357
358
359  ***************
360  *** 5545,5552 ****
361  --- 5545,5555 ----
362    TCL_H_PLACES_D="$d/generic \
363         $d/unix \
364         $d/include/tcl$TCL_HI_VERS \
365  +      $d/include/tcl$TCL_HI_VERS/generic \
366         $d/include/tcl$TCL_VERS \
367  +      $d/include/tcl$TCL_VERS/generic \
368         $d/include/tcl$TCL_ALT_VERS \
369  +      $d/include/tcl$TCL_ALT_VERS/generic \
370         $d/include \
371         /usr/local/include \
372         "
373  ***************
374  *** 6043,6049 ****
375    if test "${with_tk+set}" = set; then :
376      withval=$with_tk; d=$withval
377    else
378  !    d=""
379    fi
380
381
382  --- 6046,6052 ----
383    if test "${with_tk+set}" = set; then :
384      withval=$with_tk; d=$withval
385    else
386  !    d="/usr"
387    fi
388
389
390  ***************
```

```
391  *** 6138,6145 ****
392  --- 6141,6151 ----
393                    $d/../include/tk$TK_HI_VERS \
394                    $d/../include/tk$TK_OLD_VERS \
395          $d/include/tk$TK_VERS \
396  +       $d/include/tk$TK_VERS/generic \
397          $d/include/tk$TK_HI_VERS \
398  +       $d/include/tk$TK_HI_VERS/generic \
399          $d/include/tk$TK_OLD_VERS \
400  +       $d/include/tk$TK_OLD_VERS/generic \
401                    $d/include"
402    TK_H_PLACES=" \
403          ../include \
404  *** ns-allinone-2.35/ns-2.35/conf/configure.in.tcl.orig
        2015-07-21 03:06:55.594503300 +0900
405  --- ns-allinone-2.35/ns-2.35/conf/configure.in.tcl   2015-07-21
        03:08:34.978028700 +0900
406  ***************
407  *** 1,7 ****
408    dnl autoconf rules to find tcl
409    dnl $Header: /cvsroot/nsnam/conf/configure.in.tcl,v 1.52
        2009/12/30 22:05:31 tom_henderson Exp $ (LBL)
410
411  ! AC_ARG_WITH(tcl,  --with-tcl=path specify a pathname for tcl,
        d=$withval, d="")
412
413    dnl cant easily escape brackets in M4/autoconf-- must use
        quadigraphs below
414    AC_ARG_WITH(tcl-ver, --with-tcl-ver=path specify the version
        of tcl/tk, TCL_VERS=$withval, TCL_VERS=`echo "puts @<:@info
        patchlevel@:>@" | tclsh`)
415  --- 1,7 ----
416    dnl autoconf rules to find tcl
417    dnl $Header: /cvsroot/nsnam/conf/configure.in.tcl,v 1.52
        2009/12/30 22:05:31 tom_henderson Exp $ (LBL)
418
419  ! AC_ARG_WITH(tcl,  --with-tcl=path specify a pathname for tcl,
        d=$withval, d="/usr")
420
421    dnl cant easily escape brackets in M4/autoconf-- must use
        quadigraphs below
422    AC_ARG_WITH(tcl-ver, --with-tcl-ver=path specify the version
        of tcl/tk, TCL_VERS=$withval, TCL_VERS=`echo "puts @<:@info
        patchlevel@:>@" | tclsh`)
423  ***************
424  *** 101,108 ****
425  --- 101,111 ----
426    TCL_H_PLACES_D="$d/generic \
427          $d/unix \
428          $d/include/tcl$TCL_HI_VERS \
429  +       $d/include/tcl$TCL_HI_VERS/generic \
430          $d/include/tcl$TCL_VERS \
431  +       $d/include/tcl$TCL_VERS/generic \
432          $d/include/tcl$TCL_ALT_VERS \
433  +       $d/include/tcl$TCL_ALT_VERS/generic \
434          $d/include \
```

```
435          /usr/local/include \
436          "
437 *** ns-allinone-2.35/ns-2.35/conf/configure.in.tk.orig
         2015-07-21 03:47:27.760621900 +0900
438 --- ns-allinone-2.35/ns-2.35/conf/configure.in.tk 2015-07-21
         03:48:05.238142900 +0900
439 ***************
440 *** 2,8 ****
441   dnl $Header: /cvsroot/nsnam/conf/configure.in.tk,v 1.40
         2009/12/30 22:05:32 tom_henderson Exp $ (LBL)
442
443
444 ! AC_ARG_WITH(tk, --with-tk=path  specify a pathname for tk, d=
         $withval, d="")
445
446   dnl library version defaults to 8.0
447   AC_ARG_WITH(tk-ver, --with-tk-ver=path specify the version of
         tcl/tk, TK_VERS=$withval, TK_VERS=$TCL_VERS)
448 --- 2,8 ----
449   dnl $Header: /cvsroot/nsnam/conf/configure.in.tk,v 1.40
         2009/12/30 22:05:32 tom_henderson Exp $ (LBL)
450
451
452 ! AC_ARG_WITH(tk, --with-tk=path  specify a pathname for tk, d=
         $withval, d="/usr")
453
454   dnl library version defaults to 8.0
455   AC_ARG_WITH(tk-ver, --with-tk-ver=path specify the version of
         tcl/tk, TK_VERS=$withval, TK_VERS=$TCL_VERS)
456 ***************
457 *** 92,99 ****
458 --- 92,102 ----
459                   $d/../include/tk$TK_HI_VERS \
460                   $d/../include/tk$TK_OLD_VERS \
461       $d/include/tk$TK_VERS \
462 +     $d/include/tk$TK_VERS/generic \
463       $d/include/tk$TK_HI_VERS \
464 +     $d/include/tk$TK_HI_VERS/generic \
465       $d/include/tk$TK_OLD_VERS \
466 +     $d/include/tk$TK_OLD_VERS/generic \
467                   $d/include"
468   TK_H_PLACES=" \
469       ../include \
470 *** ns-allinone-2.35/nam-1.15/configure.orig   2015-07-21
         03:20:57.794244500 +0900
471 --- ns-allinone-2.35/nam-1.15/configure 2015-07-21
         03:47:09.294121000 +0900
472 ***************
473 *** 5642,5648 ****
474   if test "${with_tcl+set}" = set; then :
475     withval=$with_tcl; d=$withval
476   else
477 !   d=""
478   fi
479
480
```

```
481   --- 5642,5648 ----
482     if test "${with_tcl+set}" = set; then :
483       withval=$with_tcl; d=$withval
484     else
485   !   d="/usr"
486     fi
487
488
489   ***************
490   *** 5746,5753 ****
491   --- 5746,5756 ----
492     TCL_H_PLACES_D="$d/generic \
493         $d/unix \
494         $d/include/tcl$TCL_HI_VERS \
495   +     $d/include/tcl$TCL_HI_VERS/generic \
496         $d/include/tcl$TCL_VERS \
497   +     $d/include/tcl$TCL_VERS/generic \
498         $d/include/tcl$TCL_ALT_VERS \
499   +     $d/include/tcl$TCL_ALT_VERS/generic \
500         $d/include \
501         /usr/local/include \
502         "
503   ***************
504   *** 6244,6250 ****
505     if test "${with_tk+set}" = set; then :
506       withval=$with_tk; d=$withval
507     else
508   !   d=""
509     fi
510
511
512   --- 6247,6253 ----
513     if test "${with_tk+set}" = set; then :
514       withval=$with_tk; d=$withval
515     else
516   !   d="/usr"
517     fi
518
519
520   ***************
521   *** 6339,6346 ****
522   --- 6342,6352 ----
523                     $d/../include/tk$TK_HI_VERS \
524                     $d/../include/tk$TK_OLD_VERS \
525         $d/include/tk$TK_VERS \
526   +     $d/include/tk$TK_VERS/generic \
527         $d/include/tk$TK_HI_VERS \
528   +     $d/include/tk$TK_HI_VERS/generic \
529         $d/include/tk$TK_OLD_VERS \
530   +     $d/include/tk$TK_OLD_VERS/generic \
531                     $d/include"
532     TK_H_PLACES=" \
533         ../include \
534   *** ns-allinone-2.35/nam-1.15/conf/configure.in.tcl.orig
          2015-07-21 03:06:21.060265100 +0900
535   --- ns-allinone-2.35/nam-1.15/conf/configure.in.tcl 2015-07-21
```

```
         03:05:07.416254200 +0900
536 ****************
537 *** 1,7 ****
538   dnl autoconf rules to find tcl
539   dnl $Header: /cvsroot/nsnam/conf/configure.in.tcl,v 1.52
        2009/12/30 22:05:31 tom_henderson Exp $ (LBL)
540
541 ! AC_ARG_WITH(tcl,  --with-tcl=path specify a pathname for tcl,
        d=$withval, d="")
542
543   dnl cant easily escape brackets in M4/autoconf-- must use
        quadigraphs below
544   AC_ARG_WITH(tcl-ver, --with-tcl-ver=path specify the version
        of tcl/tk, TCL_VERS=$withval, TCL_VERS=`echo "puts @<:@info
        patchlevel@:>@" | tclsh`)
545 --- 1,7 ----
546   dnl autoconf rules to find tcl
547   dnl $Header: /cvsroot/nsnam/conf/configure.in.tcl,v 1.52
        2009/12/30 22:05:31 tom_henderson Exp $ (LBL)
548
549 ! AC_ARG_WITH(tcl,  --with-tcl=path specify a pathname for tcl,
        d=$withval, d="/usr")
550
551   dnl cant easily escape brackets in M4/autoconf-- must use
        quadigraphs below
552   AC_ARG_WITH(tcl-ver, --with-tcl-ver=path specify the version
        of tcl/tk, TCL_VERS=$withval, TCL_VERS=`echo "puts @<:@info
        patchlevel@:>@" | tclsh`)
553 ****************
554 *** 101,108 ****
555 --- 101,111 ----
556   TCL_H_PLACES_D="$d/generic \
557       $d/unix \
558       $d/include/tcl$TCL_HI_VERS \
559 +     $d/include/tcl$TCL_HI_VERS/generic \
560       $d/include/tcl$TCL_VERS \
561 +     $d/include/tcl$TCL_VERS/generic \
562       $d/include/tcl$TCL_ALT_VERS \
563 +     $d/include/tcl$TCL_ALT_VERS/generic \
564       $d/include \
565       /usr/local/include \
566       "
567 *** ns-allinone-2.35/nam-1.15/conf/configure.in.tk.orig
        2015-07-21 03:45:03.745348200 +0900
568 --- ns-allinone-2.35/nam-1.15/conf/configure.in.tk   2015-07-21
        03:46:01.823637300 +0900
569 ****************
570 *** 2,8 ****
571   dnl $Header: /cvsroot/nsnam/conf/configure.in.tk,v 1.40
        2009/12/30 22:05:32 tom_henderson Exp $ (LBL)
572
573
574 ! AC_ARG_WITH(tk, --with-tk=path  specify a pathname for tk, d=
        $withval, d="")
575
576   dnl library version defaults to 8.0
```

```
577   AC_ARG_WITH(tk-ver, --with-tk-ver=path specify the version of
          tcl/tk, TK_VERS=$withval, TK_VERS=$TCL_VERS)
578   --- 2,8 ----
579   dnl $Header: /cvsroot/nsnam/conf/configure.in.tk,v 1.40
          2009/12/30 22:05:32 tom_henderson Exp $ (LBL)
580
581
582   ! AC_ARG_WITH(tk, --with-tk=path  specify a pathname for tk, d=
          $withval, d="/usr")
583
584   dnl library version defaults to 8.0
585   AC_ARG_WITH(tk-ver, --with-tk-ver=path specify the version of
          tcl/tk, TK_VERS=$withval, TK_VERS=$TCL_VERS)
586   ***************
587   *** 92,99 ****
588   --- 92,102 ----
589                     $d/../include/tk$TK_HI_VERS \
590                     $d/../include/tk$TK_OLD_VERS \
591        $d/include/tk$TK_VERS \
592   +    $d/include/tk$TK_VERS/generic \
593        $d/include/tk$TK_HI_VERS \
594   +    $d/include/tk$TK_HI_VERS/generic \
595        $d/include/tk$TK_OLD_VERS \
596   +    $d/include/tk$TK_OLD_VERS/generic \
597                     $d/include"
598   TK_H_PLACES=" \
599        ../include \
600   *** ns-allinone-2.35/ns-2.35/linkstate/ls.h.orig  2015-07-21
          03:18:51.412380300 +0900
601   --- ns-allinone-2.35/ns-2.35/linkstate/ls.h 2015-07-21
          03:19:16.196984500 +0900
602   ***************
603   *** 134,140 ****
604        return ib.second ? ib.first : baseMap::end();
605   }
606
607   !   void eraseAll() { erase(baseMap::begin(), baseMap::end()); }
608   T* findPtr(Key key) {
609        iterator it = baseMap::find(key);
610        return (it == baseMap::end()) ? (T *)NULL : &((*it).second
          );
611   --- 134,140 ----
612        return ib.second ? ib.first : baseMap::end();
613   }
614
615   !   void eraseAll() { baseMap::erase(baseMap::begin(), baseMap::
          end()); }
616   T* findPtr(Key key) {
617        iterator it = baseMap::find(key);
618        return (it == baseMap::end()) ? (T *)NULL : &((*it).second
          );
619   *** ns-allinone-2.35/ns-2.35/tcl/test/test-all-template1.orig
          2015-07-30 17:08:59.818949200 +0900
620   --- ns-allinone-2.35/ns-2.35/tcl/test/test-all-template1
          2015-07-30 17:10:12.597709600 +0900
621   ***************
```

```
622  *** 123,129 ****
623      else
624              # OLD: gzip -dc $directory/$t.gz | cmp -s -
         $datafile
625              # Deleted because it is not supported in Mac OS X,
         for "cmp".
626  !      gzip -dc $directory/$t.gz > temp.randsgz
627              cmp -s temp.randsgz $datafile
628          if [ $? = 0 ]; then
629        echo Test output agrees with reference output
630  --- 123,129 ----
631      else
632              # OLD: gzip -dc $directory/$t.gz | cmp -s -
         $datafile
633              # Deleted because it is not supported in Mac OS X,
         for "cmp".
634  !      gzip -dc $directory/$t.gz | perl -ne 'print $_;' > temp.
         randsgz
635              cmp -s temp.randsgz $datafile
636          if [ $? = 0 ]; then
637        echo Test output agrees with reference output
638  *** ns-allinone-2.35/ns-2.35/bin/raw2xg.orig   2015-07-30
         17:59:08.355943800 +0900
639  --- ns-allinone-2.35/ns-2.35/bin/raw2xg 2015-07-30
         18:13:52.090026100 +0900
640  ***************
641  *** 1,7 ****
642    eval 'exec perl -S $0 ${1+"$@"}'     # -*-perl-*-
643        if 0;
644
645  ! require 5.001;
646
647    ($progname) = ($0 =~ m!([^/]+)$!);
648    sub usage {
649  --- 1,8 ----
650    eval 'exec perl -S $0 ${1+"$@"}'     # -*-perl-*-
651        if 0;
652
653  ! require 5.12.0;
654  ! use Getopt::Std;
655
656    ($progname) = ($0 =~ m!([^/]+)$!);
657    sub usage {
658  ***************
659  *** 34,41 ****
660    };
661    $usage = "usage: $progname [-a] [trace files...]\n";
662
663  ! require 'getopts.pl';
664  ! &Getopts('acefgds:m:n:qrplvt:xy') || usage;
665
666    $c = 0;
667    @p = @pc = @pg = @a = @ac = @ae = @d = @lu = @ld = ();
668  --- 35,43 ----
669    };
670    $usage = "usage: $progname [-a] [trace files...]\n";
```

```
671
672  ! #require 'getopts.pl';
673  ! #&Getopts('acefgds:m:n:qrplvt:xy') || usage;
674  ! getopts('acefgds:m:n:qrplvt:xy') || usage;
675
676  $c = 0;
677  @p = @pc = @pg = @a = @ac = @ae = @d = @lu = @ld = ();
```

CHAPTER 5

BASICS OF PROTOCOL SIMULATION USING NS2

To simulate a network protocol in NS2, we must write simulation scripts in Tcl for encoding the protocol that NS2 can invoke, and also for setting up a few necessary aspects of simulation such as nodes, links, traffic generations, etc. The focus of this chapter is the development of such Tcl scripts. After a simulation experiment, we can analyze the network/protocol performance by extracting data values from the trace output file produced during the protocol execution in the NS2 engine. To extract relevant values, AWK scripting is required, which is also discussed in this chapter. Finally, the data extracted need to be displayed as a plot using Gnuplot, which is discussed in the last section of this chapter. Data extraction and graph plotting are commonly called post-processing tasks, whereas writing the simulation script in Tcl is known as a pre-processing task.

5.1 Tcl

Tool Command Language (Tcl) is a powerful and open source programming language. It can be used for wide range of applications, including web and desktop application development. It is also used for network programming, system administration, embedded software development, testing, general purpose programming, and database programming. It is an interpreted language and supports cross-platform development. Tcl is maintained by a large community of developers worldwide. There is a core team called the Tcl Core Team (TCT) that manages the evolution of Tcl and ensures quality. The commands written are interpreted at run time by the Tcl interpreter. Everything in Tcl is a command like Unix-Shell script and it has a rich set of built-in commands that are needed for any programming language.

5.2 Program Execution

Programs may be executed either by writing a script file and then executing it, or by an interactive manner in which the commands are issued one by one in the Tcl shell. Let us first understand one of the simplest scripts. As is usually done, let us start with the classic "Hello World" program.

Listing 5.1: A hello world program (helloWorld.tcl)

```
1 # My first TCL script
2 puts HelloWorld
3 puts "Hello World"
4 puts {Hello World}
5 puts Hello World ;   #produces error
```

You may write the script given in Listing 5.1 (excluding line numbers) and save in a text file with the extension .tcl (helloWorld.tcl). Run it by issuing either of the following commands in a Linux terminal.

    $ tclsh helloWorld.tcl

or

    $ ns helloWorld.tcl

Output
HelloWorld
Helo World
Helo World
Error!!!

Program Explanation

Line 1: # is used to put a comment before a command.

Line 2: "puts" prints a string.

Line 3: "puts" prints a string with several words separated by a space but enclosed in double quotes (" ").

Line 4: "puts" prints a string with several words separated by a space and enclosed in braces ({ }). The double quotes and braces are used to groups things (strings in this case). The grouping mechanism is discussed in more detail in Section 5.3.2.

Line 5: "puts" cannot print a string with two words separated by a space; it produces an error; # is used to put a comment after a command in the same line. In other words, all statements in Tcl are written on separate lines. However, two statements can be written in same the line with a ; (semicolon) separator.

The general syntax to invoke a Tcl script from a Linux shell command prompt is

Syntax: `tclsh [<filename> <arg0> <arg1> ···]`

A Tcl program may be executed in two modes, batch or interactive. The above syntax represents a batch mode execution. It is not mandatory to write Tcl statements in a script file. All the commands available in a script file can also be issued in the Tcl shell prompt one by one to get the required result. This is called *interactive mode* execution. To illustrate this, let us perform the execution of the same program using the above syntax in an interactive manner.

Step 1: open a Linux terminal (e.g., Bash shell)

Step 2: issue the following command

    tclsh or equivalently **ns**

This would take you to the Tcl shell with a prompt as % where any Tcl commands may be issued. Now issue the following Tcl commands and observe the outputs.

 % `puts`

output: wrong # args: should be "puts <-nonewline> <channelId> string"

 % `puts HelloWorld`

output: HelloWorld

 % `puts "Hello World"`

output: Hello World

 % `puts {Hello World}`

output: Hello World

 % `puts Hello World`

output: an Error!!!

  % `puts -nonewline Hello`

  % `puts World`

output: HelloWorld

To read a value from a standard input device to a variable, use the following code.

 % `gets stdin var`

 % `puts $var`

5.3 Basic Programming Constructs

To learn any programming language, it is essential to know the built-in constructs available. Tcl does not require declaring a variable before its use (typeless). It stores everything as strings and interprets each based on the use.

5.3.1　Variables

```
% set intVar 10
% set stringVar "Tcl String Variable"
```
These variables are referenced by prefixing the variable name with a $ symbol.
```
% puts $intVar
```
output: 10
```
% puts $stringVar
```
output: Tcl String Variable
```
% puts "Int Var=$int Var, String Var = $stringVar"
```
output: Int Var = 10, String Var = Tcl String Variable.

To access the value of a variable, a $ sign is prefixed. But in some situations, it is not required. For example:
```
% incr intVar
% incr intVar -1
```
The first command increments the variable inVar and the second command decrements it. The $ sign tells Tcl to use the value of the variable. Tcl supports passing data to subroutines either by name or by value. Commands that don't change the contents of a variable usually have their arguments passed by value (like puts). Commands that need to change the value of the variable must have the variable passed by name (incr in this case). It is also possible to assign the results of a function call to a variable. Consider the example below.
```
% set a [expr 2.5*10]
% % puts $a
```
output: 25
The first command sets the value of "a" as the result of multiplication evaluated by calling the expr function with the parameter $2.5 * 10$. Tcl interprets the square brackets as delimiters for a nested command: it executes the command inside the square brackets and assigns the result to "a".

Now consider the following example given in Listing 5.2 that produces a more interesting output.

Listing 5.2: Variable manipulation in Tcl (variables.tcl)

```
1  # Variable initialization
2  set x "20+5"
3  set y "15"
4  set z $x$y
5  set t [expr $x$y]
6  puts $z
7  puts $t
8  unset z
9  puts $z
```

When executed the output result produced is as follows.

Output
20+515
535
can't read "z": no such variable
while executing
"puts z"
(file "variables.tcl" line 8)

The variable z is the string concatenation of x and y. Therefore line 6 produces 20+515. But t is evaluated numerically and the addition of two numbers (20+515=535) is produced in line 7. Line 8 clears the variable z. As a result, an error is shown when line 9 is interpreted. The general syntax of a set command is as follows.

```
% set   <var>  <value>
```

- A value is assigned to the variable var.

- A variable that contains only alphanumeric characters and no parentheses is called a scalar variable.

- If the variable has the form var(index), then it is a member of an associative array.

- A set command assigns a new value to the variable.

5.3.2 Grouping mechanism

Three types of grouping mechanisms are available in TCL:

- double quote (" ")

- double braces ({ })

- double square brackets ([ ])

The evaluation of a command in Tcl is done in two phases. The first phase is a *single pass of substitutions*. The second phase is the *evaluation of the resulting command*.

```
% set var "Hello World"
% puts $var
```

$var in the above command is substituted with the content of the variable var in the first phase, and the command is then executed in the second phase.

> Putting a string inside double quotes allows substitutions to occur within the quotations.

Table 5.1: Escape sequences

Backlash Sequence	Output
\a	Audible Bell
\b	Backspace
\f	Form Feed (clear screen)
\n	New Line
\r	Carriage Return
\t	Tab
\v	Vertical Tab
\0dd	Octal Value. d is a digit from 0 to 7
\uHHHH	This represents a 16-bit Unicode character. H is any hex digit 0 to 9, A to F, a to f
\xHH	Hex Value, H is a hex digit 0 to 9, A to F, a to f

> Grouping words using braces **disables** substitution within the braces.

However, the backslash (\) may be used to disable substitution for the single character immediately following the backslash. Also some predefined "Backslash Sequence" strings are replaced with specified values, as given in Table 5.1. Now consider the commands given below.

```
% set x 10
% puts "value of x = $x"      ;# Prints value of x
% puts "value of x = \$x"     ;# Prints a literal $x instead of the value of x
% puts Hello\World            ;# Prints 'Hello World'
% set var kid;
% puts "One $var!"            ;# Prints 'One kid'
% puts "Many $vars!"          ;#can't read "vars": no such variable
% puts "Many $var\s!"         ;#Many Kids!
% puts "Hello ${var}s!"       ;#Hello Kids!
% puts {Hello $var}           ;#Hello $var
```

Contents within braces are passed to a command exactly as written. However, backslash (\) is processed within braces at the end of a line and acts as a continuation character. An escape character is a character commonly used in a programming language that invokes an alternative interpretation on subsequent characters in a character sequence. Examples of Tcl escape sequences are provided in Table 5.1.

```
% puts Hello\World!               ;#Hello World!
```

In the above command backslash escapes the new line (both words printed in same line). Similarly, consider the following command.

```
% puts {no substitutions for \n \r \x0a \f \v}
```

Output: no substitutions for \n \r \x0a \f \v

The result of a command is obtained by placing the command inside square brackets ([ ]) like the return value of a function in C , or a back quote (') in shell programming. The string within the square brackets is evaluated as a command by the interpreter, and the square bracketed string is replaced by the result of the command. Consider the following commands.

% *set var 21*

% *puts [expr $var + 1]*   *;# Prints 22*

The above command evaluates the expression inside the square brackets, then it prints the value.

> Strings grouped within square brackets are substituted by the result of commands within the brackets.

5.3.3   Mathematical expressions

The command *expr* is used to handle mathematical and logical expressions. A few examples of mathematical expression commands are the following.

```
% expr 5 * 2 + 7
% expr "5 * 2 + 7"
% expr 5 * 2 + 7
```

All of the above produce the same output. In the case of logical expressions, any "non-zero value, TRUE, true, YES, yes" are treated as logical *true* values and similarly "0, FALSE, false, NO, no" are treated as logical *false* values.

```
% expr true && false 0
% expr true || false 1
```

The mathematical and logical operators supported are provided in Table 5.2. Similarly, there are some built-in mathematical functions available. These are given in Table 5.3. The use of the built-in functions with the command expression is illustrated in the following examples.

Table 5.2: Mathematical and logical operators

Operators	Usage
$-$, $+$	unary minus, plus
*, /, %, +, $-$	binary multiplication, division, remainder, Addition, subtraction
<<, >>	bit shift left, right
<, >, =, <=, >=	less than, greater than, equal to, less than or equal to, greater than or equal to
&, ^, |, ~	bit-wise AND, exclusive OR, OR, negation
&&, ||, !	logical AND, OR, negation
$x?y : z$	if x is non-zero, then y. Otherwise z.

Table 5.3: Mathematical functions

abs(x)	cosh(x)	log(x)	sqrt(x)
acos(x)	double(x)	log10(x)	srand(x)
asin(x)	exp(x)	pow(x,y)	tan(x)
atan(x)	floor(x)	rand(x)	tanh(x)
atan2(x)	fmod(x)	round(x)	wide(x)
ceil(x)	hypot(x,y)	sin(x)	cos(x)
int(x)	sinh(x)		

```
% expr log10(10)
```
Output: 1.0
```
% expr abs(-10)
```
Output: 10

5.3.4   Control statements

Like other languages, Tcl also provides branching and looping constructs.

Branching

As is the case with many other languages, two types of branching commands are available in Tcl.

i) `if elseif else`

ii) `switch`

Syntax:
It can be written in a single line
```
if   {expr1}   {body1}  {elseif}  {expr2}   {body2}  ... else
{bodyN}
```
or in multiple lines

```
if {expr1} {
   body1
} elseif {expr2} {
   body2
} ...
} else {
   bodyN
}
```

where *elseif* and *else* commands are optional.
Listing 5.3 demonstrates the program written with single line syntax, whereas Listing 5.4 demonstrates the program written with multiple lines syntax.

Listing 5.3: Single line if-else

```
1  if {$x > 0} {puts "$x is Positive"} else {puts "$x is Negative"
    }
```

Listing 5.4: Multiple line if-else

```
1  if {$x > 0} {
2      puts "$x is Positive"
3  } else {
4      puts "$x is Negative"
5  }
```

As can be seen from Listings 5.3 and 5.4, the if-else statements may be written in two different ways.

Listing 5.5 demonstrates a multi-way branching with if-elseif-else statements. The program prints a specific weekday according to the value of x.

Listing 5.5: Multi-way branching using if-elseif-else

```
1  puts "Enter a +ve integer (0-6)"
2  gets stdin x
3  if {$x == 0} {
4      puts "Sunday"
5  } elseif {$x == 1} {
6      puts Monday
7  } elseif {$x == 2} {
8      puts Tuesday
9  } elseif {$x == 3} {
10     puts Wednesday
11 } elseif {$x == 4} {
12     puts Thursday
13 } elseif {$x == 5} {
14     puts Friday
15 } elseif {$x == 6} {
16     puts Saturday
17 } else {
18     puts "Invalid Input"
19 }
```

Branching with switch

The switch command, as in other languages, allows choosing one of several options. Unlike many other languages, *switch* may be performed either on strings or on integers, which that makes it more flexible. Further, it allows us to compare with more than one acceptable pattern separated with a "-."
Syntax:

```
switch $string {
        pattern1 { body1}
        pattern2 {body2}
        ...
        patternN {bodyN}
        default {body}
}
```

Here string is normally the variable that is to be compared with pattern1, pattern2, etc. If the string value matches any pattern, then the code within the body associated with the corresponding pattern is executed. In one execution, only one pattern is matched.

If the string does not match any of the patterns, then the default part given in the last line of the program is executed. If it does not contain a default part, and none of the patterns match the string value in a particular execution, then the switch command returns an empty string.

Listing 5.6: Multi-way branching with switch (switch1.tcl)

```
1  puts "Enter a weekday (sunday-saturday)"
2  gets stdin val
3
4  switch $val {
5     sunday {
6        puts 0
7     }
8     monday {
9        puts 1
10    }
11    tuesday {
12       puts 2
13    }
14    Wednesday {
15       puts 3
16    }
17    thursday {
18       puts 4
19    }
20    friday {
21       puts 5
22    }
23    saturday {
24       puts 6
25    }
26    default {
27       puts "Invalid Match"
28    }
29 }
```

In Listing 5.6, an integer value (0 to 6) is printed for a given weekday. Another form of switch is given in Listing 5.7, where two patterns can match a given value. That is, if the input is either 1 or Jan, it matches case 1 and prints January.

Listing 5.7: Multi-way branching with switch (switch2.tcl)

```
1  puts {Enter an integer or month name }
2  gets stdin val
3  switch $val {
4     1 - Jan {
5        puts January
```

```
6       }
7     2 - Feb {
8           puts February
9       }
10        . . .
11  }
```

Iteration

Tcl has three loop constructs that can be used to execute the code iteratively. These are *for*, *while*, and *foreach*.
for loop:

Syntax: for   {<start>}   {<test>}   {<next>}   {<body>}

During execution, first the *start* code is executed once; this is called the initialization step. Next the *test* is evaluated as an expression. If the *test* evaluates to true, then the *body* is executed; otherwise the loop is exite, and finally, the *next* code is executed. After executing the *next* argument, the interpreter loops back to the *test* and repeats the process until the *test* evaluates to true. If the *test* evaluates to false, then the loop would be exited.

As a simple illustration of the *for* loop, see Listing 5.8, which prints 0 to 9.

Listing 5.8: loop program

```
1  for {set i 0} {$i < 10} {incr i} {
2      puts $i
3  }
```

while loop:

Syntax: while <test>   <body>

The *while* command evaluates *test* as an expression. If *test* is evaluated to true, the code in *body* is executed. After the code in *body* has been executed, *test* is evaluated again. This process is repeated until the *test* expression becomes false, after which it exits from the loop.

An example of the use of the *while* loop is provided in Listing 5.9, which prints 0 to 9.

Listing 5.9: while loop

```
1  set i 0
2  while {$i  < 10} {
3      puts $i
4  set i [expr $i + 1]
5  }
```

break and continue:

These two commands are used inside the loop. The *break* command exits the enclosing loop from the point of invocation, while the *continue* command repeats execution of the loop from the beginning without executing the commands written in the body of the loop after the point of invocation. See Listing 5.10 for an illustration of the *break* and *continue* commands.

Listing 5.10: break and continue commands

```
1  set x 5
2  while {$x < 10} {
3      set x [expr $x + 1]
4      if {$x > 8} break
5      if {$x > 6} continue
6      puts "x inside the loop is $x"
7  }
8  puts "x outside the loop is $x"
```

The output of Listing 5.10 is
x inside the loop is 6
x outside the loop is 9

foreach loop:

Syntax: foreach <varnames> < {list}> <body>

Listing 5.11: foreach loop

```
1  set observations {Bhubaneswar 35 49 Delhi 32 45 Mumbai 18 30}
2  foreach {town Tmin Tmax} $observations {
3      set Tavg [expr ($Tmin+$Tmax)/2.0]
4      puts "$town $Tavg"
5  }
```

The output of Listing 5.11 is
Bhubaneswar 42.0
Delhi 38.5
Mumbai 24.0

Test/Viva Questions

a) Write a Tcl command to concatenate two strings.

b) Write a Tcl expression for evaluating an expression of the form $\frac{x^2}{y^3}$ for given values of x and y.

c) When should one use a $ sign before a variable name?

d) Differentiate between different grouping techniques.

e) What is a backslash sequence?

f) Distinguish between the break and continue commands.

g) What are the different looping commands?

h) What are the different grouping schemes?

i) List various relational operators supported in Tcl.

j) What is the syntax of the *foreach* loop? Illustrate your answer using an example.

Programming Assignments

1) Write a Tcl program to find the largest among three given numbers.

2) Write a Tcl program to print the grade of a student based on the mark secured by him/her using the $if - elseif - else$ command. The grading scheme is
   35: 'F'
   35 - 49: 'D'
   50 - 59: 'C'
   60 - 74: 'B'
   75 - 89: 'A'
   90 - 100: 'Ex'

3) Write a Tcl program to print the multiplication table for a given number using the for loop.

4) Write a Tcl program to find the factorial of a number using the while loop.

5) Write a Tcl program to test whether a given natural number is prime or not using the while loop.

6) For a given number of lines in the output, write a Tcl program to print the following pattern using the while loop.
   ```
   *
   * *
   * * *
   * * * *
   ```

7) For a given number of lines in the output, write a Tcl program to print the following pattern using the for loop.
   ```
   * * * *
   * * *
   * *
   *
   *
   ```

8) Write a Tcl program to find the sum of the digits of a given integer number.

9) Write a Tcl program to print the age of a person given date of birth and current date.

10) Write a Tcl program to generate the Fibonacci series numbers up to a given number of terms.

5.4   Arrays

An array in Tcl can be used to store multiple elements. Actually, an array in Tcl is implemented as an associative map. An associative map is a structure in which an element in an array is accessed based on its key value. That is, it associates an array index (key) with an array element. Unlike C/C++ arrays, the Tcl array index can be a string.

Listing 5.12: Storing and retrieving an element from an array (array1.tcl)

```
1   # Numeric indexing
2    set myArray(0) 1
3    set myArray(1) 65
4   # String indexing
5    set Link(pktsize) 512
6    set Link(protocol) "IEEE 802.11"
7   # Printing Array values
8    puts $myArray(0)
9    puts $myArray(1)
10   puts $Link(pktsize)
11   puts $Link(protocol)
```

Output of Listing 5.12:
1
65
512
IEEE 802.11

Listing 5.13: Other ways of initialization (array2.tcl)

```
1   # Other methods of initializing
2   set val1 5
3   set myArray($val1) 500
4   set val2 key
5   set myArray($val2) "abc xyz"
6   puts $myArray(5)
7   puts $myArray(key)
8   puts $myArray($val1)
9   puts $myArray($val2)
```

Output of Listing 5.13:
500
abc xyz
500
abc xyz

A few other commands are available for array operations. Some of commonly used ones are given below.
```
% array exists myArray
```
Used to test if an array exists. Returns 1 if an array exists, otherwise returns 0.

```
% array unset myArray
```
Used to delete an array.
```
% info exists array(key)
```
To test if a particular key of the array exists
```
% array unset myArray prefix,*
```
Used to delete one or more keys, specifying the pattern.
```
% array get myArray
```
Converts an array to a list (list is discussed in the next subsection.
```
% array get myArray prefix,*
```
Used to get specific elements of an array by adding a pattern to match keys.
```
% array set myArray "1 first 2 second"
```
Used to set one or more keys in an array from the name-value pairs.

In this example, the array my Array will store first in index 1 and second in index 2. To check this, print the array content using the following command.
```
% puts $myArray(1); puts $myArray(2)
```

```
% array names myArray
```
This command prints the key names (index) of an array.
```
% array names myArray prefix,*
```
Used to get only a subset of names. This will return only the keys which start with the word prefix.
```
% array size myArray
```
Returns the number of key-value mappings.

In Tcl, there is no in-built provision for storing and manipulating multi-dimensional arrays. However, we can simulate a two-dimensional array by any of the following naming conventions.
```
% set a(0,0) 5              ;# set element 0,0 to 5
% set a(0.0) 5
% set a(0:0) 5
```

The array can be accessed and manipulated in a similar manner. Listing 5.14 demonstrates reading and printing of a two-dimensional array.

Listing 5.14: Multidimensional array

```
1 puts "Enter 4 elements (press Enter key after each element)"
2 for {set i 0} {$i < 2} {incr i} {
3   for {set j 0} {$j < 2} {incr j} {
4     gets stdin A($i,$j)
5   }
6 }
7 puts "The Matrix Elements are:"
8 for {set i 0} {$i < 2} {incr i} {
9   for {set j 0} {$j < 2} {incr j} {
10    puts -nonewline "$A($i,$j) \t"
11  }
```

```
12    puts \n
13    }
```

5.5   Lists

A list is a basic Tcl data structure that holds an ordered sequence of values.
Lists can be created in several different ways as follows.

```
% set myList [list one two three four]
```
or
```
% set myList "one two three four"
```
or
```
% set myList  one two three four
```

All of the above commands have the same effect. As can be seen, there is
no visible difference between a string and a list.

To print the list
```
% puts $myList
```

A list may also contain another list, and this can be handled as follows.
```
% set nestedList [list one two three [list 1 2 3]];
```

There are a few commands available to handle lists as follows.
```
% llength [list hello how are you]     ;# 4,
```
Above command is used to get the number of elements in the list.
```
% lindex $myList 0     ;# one, i.e., first element of myList
% lindex $myList end   ;# four, i.e., last element of myList
% lindex $myList end-2     ;# two,
```
Above command is used to retrieve an element specified as the third element
from the end of myList.
```
% lindex $nestedList 2 1     ;# 2
```
The above command is used to retrieve an element from the nestedList, i.e.,
the first element of the second list in nestedList.
```
% puts [lrange $myList 1 3]     ;# one two three
```
The above command is used to get a range (1 to 3) of elements from myList.
```
% lappend myList five six     ;# string 'five' and 'six' are appended
```
to myList
```
% puts $myList     ;# one two three four five six,
```
lappend command appends elements (i.e., five six) at the end of myList
```
% linsert $myList 1 hello     ;# string 'hello' is inserted after the
```
first element (one) in myList
```
% puts $myList     ;# one hello two three four five six,
```
This inserts one or more elements starting from the position specified by the

first argument, i.e., after the first element in this example.

```
% set list1 [list 1 2 3 4]
% lset list1 0 5        ;# replaces '1' with '5' in list1
% puts list1            ;# 5 2 3 4
```
The *lset* command replaces the element specified in the first argument with the value specified in the second argument.
```
% set list1 [ list 3 5 1 9]
% puts [lsort $list1]           ;# 1 3 5 9
```
lsort sorts the elements of list1 in ascending order.
```
% puts [lsort -decreasing $list1]           ;# 9 5 3 1
```
-decreasing argument to *lsort* is used to sort the elements of list1 in descending order.

By default, the list elements are arranged as text values. Therefore the result of the following command produces undesired outputs.
```
% set list1[list 10 5 25 6]
% puts [lsort $list1]        ;# 10 25 5 6
```

Therefore, to avoid such situations, there are some switches available to define the type of comparisons as follows:

-ascii	compares as text (default switch)
-integer	integer comparison
-real	floating point comparison
-dictionary	lexicographical

The *lsort* command can also be used to sort the list according to the specified element of the sublists by using a switch index as follows.
```
% puts [lsort -integer -index 1 {{Alok 3}{Bijoy 1}{Kartik 2}}]
              {Bijoy 1}{Kartik 2}{Alok 3}
```

To search an element from a list, the command *lsearch* may be used.
```
% set list1 [list 1 2 3 4]
% puts [lsearch $list1 3]        ;#2
```
This command returns the index, if the element is found, otherwise it returns −1.

5.6 Dictionaries

Like arrays, dictionaries maintain a map between key and value, but unlike an array, a dictionary can be truly multi-dimensional and is treated as an object with faster execution operations than arrays.

To create a dictionary
```
% set myDict [dict create key1 val1 key2 val2]
```
To access the value
```
% dict get $myDict key2 val2
```
To create multilevel dictionaries
```
% set multiDict [ dict create 1[ dict create 2 [ dict
create 3 25]]]
```
The first key is mapped to a dictionary, the second key is also mapped to another dictionary, but the third key is mapped to a value.

To access the value
% get dict $multiDict 1 2 3 25

5.7   Procedures

In Tcl, "commands" and "keywords" are synonymous. Unlike other languages, Tcl does not have reserve/key words like if and switch, for, while, etc., but these are implemented as commands. The Tcl interpreter comes with a set of built-in commands that includes while, for, set, puts, etc.

Commands are like "functions" in other languages. Therefore you are allowed to create a new command by writing a procedure, and that works exactly like a built-in command. The *proc* command creates a new command. The syntax for the proc command is

```
% proc   <procName>  {<args>}  {<body>}
```
Invoke the procedure as follows:

```
% <procName>  <args>
```

When a *proc* is defined, it creates a new command with the name procName that takes arguments args. When the procedure procName is called, it then runs the code contained in the body. A list of arguments (may be empty) args is passed to procName. When procName is invoked, local variables with these names will be created, and the values to be passed to procName will be copied to the local variables. The body contains commands to be executed when the procedure is invoked. A body can be defined either with or without a return command. The return command returns its argument to the calling program. If there is no return, then the body will return to the caller when the last command of the body finishes the execution.

Listing 5.15: Procedure with no arguments and without any return value

```
1  proc myCommand {} {
2     puts "My first Tcl command"
3  }
4  myCommand   ;# My first Tcl command
```

Listing 5.16: Procedure with parameters and no return value

```
1  proc  myCommand {a b} {
2      puts [expr $a + $b]
3  }
4  myCommand 10  20  ;#30
```

Listing 5.17: Procedure that returns a value

```
1   proc avg {min max} {
2          return [expr ($min + $max) / 2]
3   }
4   set average [avg 10 20]
5   puts $average    ;#15
```

Listing 5.18: Generation of a random value in a range

```
1  #).*rand(1000,1)  +  a(b-a).*rand(1000,1)  +  a Random Number
       Generation
2  proc findRand {min max} {
3      set randVar [expr  ($max-$min)*rand() +$min]
4      return $randVar
5  }
6  #Obtain a uniform random value between 0 and 1
7  set randVar [findRand 0 1]
8  puts $randVar
```

Listing 5.18 produces different outputs/ results for different execution runs.

Listing 5.19: Procedure with default arguments

```
1  proc add {r1 {r2 100} {inc 1}} {
2    set sum 0
3    for {set i $r1} {$i <= $r2} {set i [expr $i + $inc] } {
4          set sum [expr $sum + $i]
5      }
6      return $sum
7  }
8  set sum1 [add   1]
9  set sum2 [ add 1 10 ]
10 set sum3 [ add   1 10 2 ]
11
12 puts $sum1 ;# 5050
13 puts $sum2 ;# 55
14 puts $sum3 ;# 25
```

The example programs given in Listings 5.15 – 5.20 illustrate passing parameters to a procedure and returning a value from a procedure. A procedure can also have default values for its arguments, as shown in Listing 5.19. This procedure may be invoked as follows:

```
% set sum [add 1] ;# 5050
% set sum [ add 1 10 ] ;# 55
% set sum [ add 1 10 2 ] ;# 25
```

Since there are default arguments, for the first call, it assumes the value for r2 and inc as 100 and 1, respectively. For the second invocation, the value of r2 is given as 10, so it will not take the default value for r2 but assumes the value of inc as 1. In the last invocation, it does not take any default value as all values are specified in the procedure call.

A *proc* can also accept a variable number of arguments if the last declared argument is the word *args*. If the last argument to a *proc* argument list is *args*, then any arguments that aren't already assigned to previous variables will be assigned to *args*.

Listing 5.20: Procedure with a variable number of arguments

```
1 proc add {val args} {
2    set resList {}
3    foreach element $args {
4       lappend resList [expr $element + $val]
5    }
6    return $resList
7 }
8 set resList [add 5 200 300 400 500]
9 puts $resList
```

The example procedure *add* in Listing 5.20 is defined with two arguments. At least one argument *must* be present when the *proc add* is invoked. By declaring *args* as the last argument, it can take a variable number of arguments. In this program five arguments are passed in line 8. The output of Listing 5.20 is 205 305 405 505.

Sometimes it is necessary within a procedure to reference a variable that has global scope, as given in Listing 5.21.

Listing 5.21: Global variable referencing

```
1 set var 20
2 proc globProc {} {
3    global var
4    set var [expr $var+10]
5    puts $var
6 }
7 globProc
8 puts $var
```

In Listing 5.21, in line 7, when the procedure is invoked, the value of global variable *var* is incremented by 10 and printed in line 5. Again in line 8 it prints the incremented value as it has been modified in line 4 inside the procedure. The output is 30 30.

5.8   File Handling

A disk file is accessed in Tcl using channels. A channel is like a C++ stream that can be a file, a network socket, a pipe, or any other channel type.

Depending on the type of channel, it can support operations for reading, writing, or both. There are three default channels — *stdin*, *stdout*, and *stderr*; *stdin* is used to read information from the standard input device (keyboard); *stdout* is used to write information to the standard output device (terminal display), and *stderr* is used to write errors to the stderr device, which is usually the terminal display.

```
% puts stderr "Error Channel"
```
output: Error Channel
```
% puts stdout "TCL Version: [info tclversion]"
```
output: TCL Version: 8.4

The following command is used to read any value from the keyboard to a variable.
```
% gets stdin val
```
input: 50   ;# entered from keyboard
output: 2   ;# number of characters read

The following command is used to write the value of a variable to the terminal display.
```
% puts stdout $val
```
output: 50   ;#value of *val* printed

5.8.1   Reading and writing files

The *open* command may be used to open a file for read/write operation. It can be invoked in any of the following three ways.

   i) using the filename

   ii) using the filename and the open mode

   iii) using the filename, open mode, and file permissions

The default file open mode is read mode, and file permissions are set as the default permission mode of the operating system when it is not specifically set by the user. The open command returns the name of the newly opened channel. Different file opening modes and their various characteristics are listed in Table 5.4.

Listing 5.22: Reading from a file: 1st Method

```
1 set f [open complex.tcl r]
2 while   {![eof $f]} {
3   gets $f line
4   puts $line
5 }
6 close $f
```

Table 5.4: File opening modes

Mode	Read	Write	Start position	Must exist	Erased
r	Yes	No	Beginning of file	Yes	No
r+	Yes	Yes	Beginning of file	Yes	No
w	No	Yes	Beginning of file	No	Yes
w+	Yes	Yes	Beginning of file	No	Yes
a	No	Yes	End of file	No	No
a+	Yes	Yes	End of file	No	No

Listing 5.22 opens a file complex.tcl in read mode, reads from it, and displays each line. If complex.tcl does not exist then the program shows an error. Listing 5.23 would cause the total contents of the file to be read one at a time and displayed. Listing 5.24 opens two files, one in read mode, the other in write mode. It reads the contents of one file and writes the same to the new file, i.e., complex1.tcl. After the program is executed, a file with the name complex1.tcl is created in the disk with the same contents as the existing file complex.tcl.

Listing 5.23: Reading from a file: 2nd method

```
1 set inF [open complex.tcl r]
2 set data [read $inF]
3 puts $data
4 close $inF
```

Listing 5.24: Program to copy contents of a file

```
1 set fp [open complex.tcl r]
2 set fp1 [open complex1.tcl w]
3 set data [read $fp]
4 puts $fp1 $data
5 close $fp
6 close $fp1
```

Test/Viva Questions

a) What is the difference between a C array and a Tcl array?

b) What is a list in Tcl?

c) What is the difference between an array and a dictionary?

d) What is the syntax of writing a procedure?

e) How can one write a procedure with default arguments?

f) How can one write a procedure with a variable number of arguments?

g) How can one access a global variable in a procedure?

h) What is the syntax to open a file in write mode?

i) What are the different file opening modes? What are their implications?

j) What is the difference between reading a file using *gets* as compared to using *read*?

Programming Assignments

1) Write a Tcl script to search an element in an array of integers.

2) Write a Tcl script to arrange the integer elements of an array in ascending order.

3) Write a Tcl script to reverse the elements of an array in place, i.e., no second array is used.

4) Write a Tcl procedure to swap values of two variables and display the swapped values.

5) Write a Tcl procedure to return the largest value of an array of values passed in to the procedure.

6) Write a Tcl procedure that returns 1 if the number passed to it is a prime number and returns 0 otherwise.

7) Write a Tcl procedure that multiplies a matrix with a vector. Both the matrix and the vector should be passed as arguments.

8) Write a Tcl program to read two compatible matrices from a file, multiply them, and write the result into another file.

9) Write a Tcl program that counts the word frequencies present in a given text file and displays the result.

10) Write a Tcl program to arrange a set of strings available in a text file in lexicographic order using an array and display the result.

5.9   Object-Oriented Tcl (OTcl)

It is expected that readers have already been exposed to object-orientation concepts and are familiar with an object-oriented programming language like C++. In this section the object-oriented constructs of Tcl are explained. **Note:** Tcl itself does not support objects, but the OTcl extension used by NS2 supports these, as mention in Chapter 3.

5.9.1   Classes and objects

Some of the special words and syntax used in the context of OTcl are as follows.

i) Class is used to register a class name.

```
class Counter          ;# Counter is a user defined class.
```

ii) instproc is used to define a member function.

```
Counter instproc increment { } {
# increment is a member function of class Counter
...
}
```

iii) init is the name of a constructor.

```
Counter instproc init { } {
# It is a constructor of class Counter
...
}
```

iv) self is equivalent to the *this* keyword of C++. That is, it represents the invoking object inside a member function or to access data members called instance variables.

v) instvar is used to declare a data member. Using this construct, the data members of a class need not be declared in advance, and can be created on demand. If one instance variable is already declared in its class or in its superclass, the variable is referenced, otherwise a new one is declared.

```
self instvar val ;# declare an instance variable 'val'
set val 10 ;# Assign 10 to 'val'
```

vi) superclass is used to specify the parent class in the case of inheritance.

```
class Kid -superclass Mom ;# Kid class is inherited from Mom class.
```

vii) Object initialization: To initialize an object the syntax is

```
set ctr [ new Counter ] ;# ctr is an object of class Counter
```

It calls the constructor to initialize the instance variables.

viii) Method invocation: To call a class method we need to write

```
$ctr increment ;# invokes the method increment of class Counter
puts [ $ctr getValue ]    ;# invokes the method getValue of class
```

Counter and prints the value returned from the member function

ix) next is used to call the parent class constructor inside a child class constructor.

```
class myNode - superclass Node
...

myNode instproc init {} {
        # invoke parent class constructor
        $self next
        $self instvar datamember1
        set datamember1 0
}
```

Unlike C++, there is no class body where all members of the class are declared/defined. That is, once the class is registered, the member functions and data members can be declared anywhere in the script with the use of the class name and self variable.

Listing 5.25: A simple class program in OTcl (counter.tcl)

```
1  #--- counter.tcl ---
2  # run: ns counter.tcl
3  # Create a Class
4   Class Counter
5
6  #Constructor for the class Counter
7   Counter  instproc init { } {
8          $self instvar val
9          set val 0
10  }
11
12  #Member function
13   Counter instproc increment { } {
14          $self instvar val
15          set val [ expr $val + 1 ]
16  }
17
18  #Member function
19   Counter instproc decrement { } {
20          $self instvar val
21          set val [ expr $val - 1]
22  }
23
24  #Member function
25   Counter instproc getValue { } {
26          $self instvar val
27          return $val
28  }
29
30  #Declare an Object of Counter Class
31   set ctr [ new Counter ]
32
33  #Calling member function increment
34   $ctr increment
35   $ctr increment
36  #Calling member function getValue
37   puts [ $ctr getValue ] ;#ouput:
38
39  #Calling member function decrement
40   $ctr decrement
41   puts [ $ctr getValue ] ;#ouput:
```

The class counter in Listing 5.25 has three member functions: the *increment* member function increments the value of the data member *val* by one, the *decrement* member function decreases the data value of *val* by one, and the *getValue* member function returns the data value of *val*. It also has a *constructor* that is used to create the object and intialize the data value of *val* to zero. In lines 32 and 33, the increment function is called twice; therefore in line 35 it produces output 2. Again in line 38 it decrements the value once due to which it prints 1 in line 32.

The example code provided in Listing 5.26 implements a complex number class. Lines 2 to 7 are used to define the constructor that initializes the member variables. Lines 9 to 15 are used to define a member procedure that adds two complex number objects and returns the result. The member procedure disp used to display the complex number is given in lines 17 to 22. Two other procedures are defined in lines 24 to 32 that return the data member values. Then three complex number objects (c1, c2, and c3) are created in lines 34 to 36. Next two objects are added and the result is stored in the third object in line 38. Lines 40 to 45 are used to display different object values.

Listing 5.26: Complex number class (complex.tcl)

```
1  #--- complex.tcl
2  # creates two complex objects and adds them
3  #run: ns complex.tcl
4  Class Complex
5
6  Complex instproc init {a b} {
7     $self instvar real
8     $self instvar img
9     set real $a
10    set img $b
11 }
12
13 Complex instproc add { a b} {
14    $self instvar real
15    $self instvar img
16    set real [expr [$a getValueReal] + [$b getValueReal] ]
17    set img [expr [$a getValueImg] + [$b getValueImg] ]
18    return $self
19 }
20
21 Complex instproc disp {} {
22    $self instvar real
23    $self instvar img
24    puts -nonewline "$real + i"
25    puts $img
26 }
27
28 Complex instproc getValueReal {} {
29    $self instvar real
30    return $real
31 }
32
33 Complex instproc getValueImg {} {
34    $self instvar img
35    return $img
36 }
37
38 set c1 [new Complex 3 4]
39 set c2 [new Complex 5 6]
40 set c3 [new Complex 0 0]
```

```
41
42 set c3 [$c3 add $c1 $c2]
43
44 puts -nonewline "c1: "
45 $c1 disp
46 puts -nonewline "c2: "
47 $c2 disp
48 puts -nonewline "c3: "
49 $c3 disp
```

5.10   AWK Scripting

AWK is normally used to extract required information from a text (ASCII) file. While using NS2, it is used to extract information from the trace files produced after executing a simulation. AWK is so named after its creators "Aho, Weinberger, and Kernighan" and was originally developed at AT&T. AWK is a data-driven scripting language suitable for text processing or data extraction and report formatting. It is interpreted line by line and produces the output in a required format. Input text files are considered to be a set of records containing different fields, i.e., each line in the file is a record (row) and each word of a line is considered as a field (column). The default column separator is white space; however, other delimiters such as a comma (,) or a tab space may also be used. The first column of the file is accessed using $1, the second column using $2, etc.

The basic syntax of an AWK command is as follows:

pattern {action(s)}

pattern specifies a condition; *action* is performed on a line in the input file if the condition in *pattern* is satisfied for that line. However, *pattern* is optional and if *pattern* is absent, then the specified action is performed for every record (line) of the input file. A pattern may be a search pattern in the form of a regular expression or special patterns such as *BEGIN* and *END*.

AWK commands can be executed either from the *command line* or by writing multiple commands in a *script file* and then the file can be executed. Normally, if we have a simple task to do, then we choose the first method, but when we have a reasonably large number of tasks to be performed on a file, then we use an AWK script file. Let us learn each of these methods one by one. Let us consider a text file (file.txt) with the following content.

```
Ashok 20-sep-1991 India
Marry 13-oct-1995 Italy
John 15-mar-1990 USA
Karim 25-Feb-1993 Pakistan
```

- **Command Line Execution**
  Issue the following commands in the command prompt.

```
        awk < file.txt '{print $2}'
```
or
```
        awk '{print $2}' file.txt
```
where *file* is the input input text file, on which the command needs to execute. As you can see, the command does not contain any pattern; therefore it will display the second field (column) of each line available in a text file called file.

Then output of the above AWK command will be

20-sep-1991

13-oct-1995

15-mar-1990

25-Feb-1993

Now let us test another command.
```
        awk '/Italy/ {print }' file.txt
```
The output for this would be

Marry 13-oct-1995 Italy

In this command a search pattern is used, and it extracts the complete line containing the pattern Italy.

- **Executing from a Script File**
  In this case write the AWK commands in a file. To execute an AWK script file, the command syntax is

```
        awk -f <awk_file><input_file>
```

where *awk_file* is the AWK script file and *input_file* is the data file on which the operations are to be performed. Now write the following command as given in Listing 5.27 in a file (myFirstAwk.sh) as follows.

Listing 5.27: First AWK script

```
1  #!/bin/bash}
2  { print $2}
```

To execute the above programs issue the following command.
```
        awk -f myFirstAwk.sh file.txt
```
This will produce the same output as above. Similarly, the AWK file may be made executable by modifying the first line of myFirstAwk.sh as follows.
```
        #!/usr/bin/awk -f
        { print $2}
```
Then change the permission mode of the script file as follows.
```
        chmod +x myFirstAwk.sh
```
Now execute as follows.
```
        ./myFirstAwk.sh file.txt
```
Again the same output is produced.

5.10.1   General structure of AWK

```
BEGIN action(s)
         A set of pattern-action(s) pair
END action(s)
```

All actions with a BEGIN pattern are executed once at the beginning of the execution of the script. Normally, the variable initializations are made in this section. The pattern-action pair between the BEGIN and END patterns is executed for each line of the input file. Finally, actions with the END pattern are executed once at the end. The END section is normally used for printing the summarized results.

5.10.2   Other AWK constructs

- **Operators**
  Operators are mostly like C operators, given as follows.
  Arithmetic: +,   −,   *,   /,   %,   <space> (string concatenation)
  Increment, decrement: ++,   − −
  Assignment: =,   + =,   − =,   * =,   / =,   % =
  Relational: ==,   <,   >,   <=,   >=,   ! =
  Regular Expression: ~ (matching),   ! (not matching)
  Logical: &&, ||,   !

- **Control Structure**
  The keywords used are
  ```
  if, while, for, break, continue, next, exit
  ```
  Syntax:
  if (condition) <actionSet1> [**else** <actionSet2>]
  while (condition) <actionSet>
  for (<initialization>; <condition>; <actions>)
  <actions>

- **Output**
  Regular display: print <item1> <item2> ...
  C Style display: printf(<format>,<item1>,<item2> ...)

- **Comment**
  A line starting with the # symbol is called a comment line.

- **Predefined variables**
  $0: entire record, $1: first column, $2: second column, ...
  NR: number of input records
  NF: number of fields in a record
  FILENAME: input file
  FS: field separator

RS: record separator

etc.

Examples

Let us examine a few example AWK programs. For all the programs, consider
a text file that contains a student's name and grades secured by that student
in various subjects in a row as follows.

Input Text File: *file.txt*

```
1001    Ashok       83   73   93
1002    Ram         30   43   67
1003    John        45   69   83
1004    Marry       43   35   59
1005    Samita      20   33   47
1006    Shibani     66   66   76
1007    Sourav      30   25   45
1008    Samyak      71   75   65
1009    Rudra       30   43   67
1010    Laila       85   83   87
```

Listing 5.28: List first division students

```awk
1  #!/bin/awk -f
2  #program to print student names scoring First division
3  # run: awk -f firstDiv.sh < file.txt
4  BEGIN {
5     lines=0;
6     print "\n Students with First Division";
7     print "=============================";
8  }
9  {
10    lines++;
11    avg=($3+$4+$5)/3;
12    if (avg >= 60) {
13          printf $2 "\t" avg "\n";
14       }
15  }
16  END {
17     print "=============*===============";
18     print lines "  lines scanned...";
19  }
```

OUTPUT:

Students with First Division

============================

Ashok 83%

John 65.6667%

Shibani 69.3333%

Samyak 70.3333%

Laila 85%

=============*===============

10 lines scanned...

Listing 5.29: List failed students

```awk
#!/bin/awk -f
#program to print failed student names.
#run awk -f failedStud.sh < file.txt
BEGIN {
    print "\nFail Students list";
    print "====================";
}
{
  if ($3 < 30 || $4 < 30 || $5 < 30) {
          printf $1 "\t" $2 "\t" "\n";
  }
}
END {
    print "=========*=========";
}
```

OUTPUT:

Fail Students list

====================

1005 Samita

1007 Sourav

=========*=========

Listing 5.30: List students having poor performance

```awk
#!/bin/awk -f
#program to print students scoring less than class average
    using associative array
BEGIN {
  print "\n List of Students having poor performance";
  print "====================";
  student=0;
  totMark=0;
}
{
  totMark += $3 + $4 + $5;
  indvAvg[$2] = ($3 + $4 + $5)/3;
  student++;
}
END {
  classAvg=totMark/(student*3);
  for (i in indvAvg)
          if(indvAvg[i] < classAvg)
                  print i "\t" indvAvg[i] "\n";

  print "--------------------\n";
  print "Class Avg: " classAvg;
}
```

OUTPUT:

List of students having poor performance

====================

Marry 45.6667

Ram 46.6667
Samita 33.3333
Sourav 33.3333
Rudra 46.6667

Class Avg: 57.9

Listing 5.31: Display file names along with their size

```
 1  #!/bin/sh
 2  #execute with: ls -l | awk -f fileSize.sh
 3  BEGIN {
 4    print "File \t Size"
 5  }
 6  {
 7    print $9, "\t", $5
 8  }
 9  END {
10    print " - DONE -"
11  }
```

The reader is encouraged to enter the given scripts in script files, execute them, and observe the output.

5.11   Gnuplot

Gnuplot is a free and open-source software used to plot graphs using data or functions. It is command driven and suitable for 2D as well as 3D plots.

Gnuplot uses the same functions as available in the math library; some of these are listed in Table 5.5.

Gnuplot contains many commands to produce and customize plots some of the useful commands are given in Table 5.6.

To start Gnuplot, issue the command Gnuplot in the OS command line, which will a result in a prompt like
gnuplot>
This is called the Gnuplot command line. A plot now can be created using either of the following two ways.

i) use the plot command in the command line, or

ii) create a script file with extension <script_file>.p, then *load* the script.

Similarly, a plot can be drawn by taking input data either from a function or from a data file.

Plot using a predefined function
Gnuplot> plot sin(x)/x
This assumes a default set of values for x and plots the value of the function on the y-axis for each x. The output produced is given in Figure 5.1.

Table 5.5: Functions usable in Gnuplot

Function	Return
abs(x)	absolute value of x, |x|
cos(x)	cosine of x
exp(x)	exponential function of x
log(x)	natural logarithm (base e) of x
log10(x)	logarithm (base 10) of x
rand(x)	pseudo random number generator
real(x)	real part of x
sin(x)	sine of x
sqrt(x)	square root of x
tan(x)	tangent of x
...	

Table 5.6: Some Gnuplot commands

Command	Meaning
Gnuplot	to start Gnuplot environment; results in a prompt like Gnuplot>
quit	exit from Gnuplot
help plot	introductory help
help <topic>	help on a topic
show all	current environment
plot <function>	2D plot of a function
splot <function>	3D plot of a function
plot <dataFile> using col_x:col_y	plots by taking data from datafile
load '<scriptFile>.p'	executes the scriptFile

Plot using a data file containing data elements
Consider the data file *scheduling.dat* containing some arbitrary contents as
follows.

```
# File Name: scheduling.dat
# Processor requirement for differnt algorithms
#Tasks      EDF         MinUtil         MaxUtil
70          5           8               10
80          6           10              11
90          7           12              12
100         10          14              15
110         10          16              18
120         11          19              22
130         12          25              23
```

| 140 | 13 | 30 | 24 |
| 150 | 13 | 34 | 26 |

Now to plot from this data file, issue the following command.

Gnuplot>   plot "scheduling.dat" using 1:2 title 'Earliest Deadline First' with lines, \
            "scheduling.dat" u 1:4 t 'Max Util First' w linespoints

The words using , title , and with may be abbreviated as u, t, and w, respectively, i.e., each word can be abbreviated to its smallest nonconflicting letter/word. Also, each line and point style has an associated number.

The above plot command plots two curves. The first curve is plotted by taking col_1 on the x-axis and col_2 on the y-axis, and the second curve is drawn using col_1 values on the x-axis and col_4 values on the y-axis. The output plot is shown in Figure 5.2.

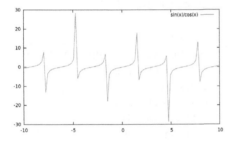
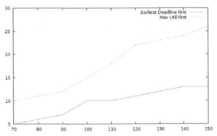

Figure 5.1: Plot of sin(x)/cos(x)   Figure 5.2: Plot using a data file

5.11.1   Customizing plots

To generate a plot, one may need to use several commands. To avoid typing multiple commands in the command line, a script file may be created containing the required commands. The script file may be reused many times without writing all these commands again and again for each plot generation. To generate a customized plot for Figure 5.2, the script file listed in Listing 5.32 may be used.

Listing 5.32: First Gnuplot script

```
1 #Gnuplot script file
2 #File Name: scheduling.p
3 #run: load 'scheduling.p'
4 set    autoscale # scale axis automatically
5 set xtic auto    # set xtics automatically
6 set ytic auto    # set ytics automatically
7 set title "Processor requirement in different Algorithms"
8 set xlabel "Processors"
9 set ylabel "Tasks"
```

```
10 plot "scheduling.dat" u 1:2 t 'Earliest Deadline First' w lines,
     \
11 "scheduling.dat" u 1:3 t 'Max Util First' with linespoints
```

To execute this script, load the script in the Gnuplot command line as follows.

Gnuplot>load 'schedule.p'

Providing a script file within single quotes (' ') is mandatory. Lines 1 and 2 are comment lines, i.e., any text written after a # symbol is considered comment. Lines 3 to 11 are used to set different plot parameters as required. Lines 12 and 13 are the plot commands. This script takes data from scheduling.dat and plots two lines with different line styles.

Plots may be displayed in any of the styles given in Table 5.7. To test

Table 5.7: Plot styles

Style	Meaning
dots	displays a small dot at each point
points	displays a small symbol at each point
lines	connects adjacent points with lines
linespoints	draws both lines and points
impulses	displays a vertical line from the x axis to each point
steps	displays the curve by joining the points in steps

different styles, issue the following command.

Gnuplot> plot sin(x) title 'points' w p, sin(x+0.5) title 'lines' w l, sin(x+1) title 'linespoints' w lp, sin(x+1.5) title 'dots' w d

which generates a plot shown in Figure 5.3.

For each of the above styles, the line width, line type, point size, and point types for a curve may be customized as given below.

```
   with <style> linewidth <lineWidth> linetype <lineType>
        pointsize <pointSize> pointtype <pointType>
```

or

```
   w <style> lw <lineWidth> lt <lineType> ps <pointSize> pt
<pointType>
```

where *style* is one of the above styles. The *lineWidth*, *lineType* and *pointSize*, *pointType* are positive integer constants. Line type 1 is the first line type used by default, line type 2 is the second line type used by default, etc. A command called test is available that may be used to generate a standard test image showing all available line styles and fill patterns. To use the test command, use the test command in Gnuplot's command line as follows

Gnuplot>test

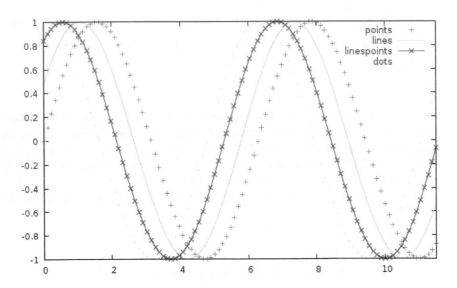

Figure 5.3: Plot using different styles

This produces a demo plot given in Figure 5.4. Some example plots for different values of *lineWidth, lineType, pointSize, and pointType* are shown in Figures 5.5 – 5.8. The following commands may be used to produce these plots.

Command for Figure 5.5:

```
Gnuplot> plot sin(x) title 'lw=1' w l lw 1, sin(x+0.5) title
'lw=2' w l lw 2, sin(x+1) title 'lw=3' w l lw 3, sin(x+1.5)
title 'lw=4' w l lw 4, sin(x+2) title 'lw=5' w l lw 5
```

Command for Figure 5.6:

```
Gnuplot>plot sin(x) title 'line type=2' w l lt 2, sin(x+0.5)
title 'line type=4' w l lt 4, sin(x+1) title 'line type=6' w l
lt 6, sin(x+1.5) title 'line type=8' w l lt 8
```

Command for Figure 5.7:

```
Gnuplot> plot sin(x) title 'point size=1' w p ps 1, sin(x+0.5)
title 'point size=2' w p ps 2, sin(x+1) title 'point size=3' w
p ps 3, sin(x+1.5) title 'point size=4' w p ps 4
```

Command for Figure 5.8:

```
Gnuplot> plot sin(x) title 'point type=2' w p pt 2, sin(x+0.5)
title 'point type=4' w p pt 4, sin(x+1) title 'point type=6' w
p pt 6, sin(x+1.5) title 'point type=8' w p pt 8
```

The command set can be used to set different plot parameters as follows.
To choose the angle type
```
    set angles [degrees|radians]
```
To draw an arrow "from point" to "to point"

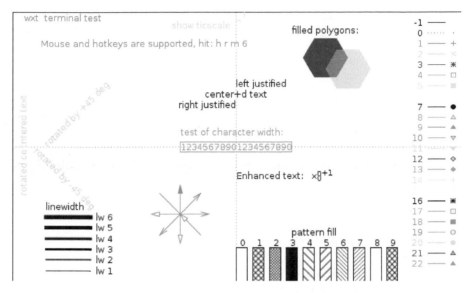

Figure 5.4: Demo plot showing customization

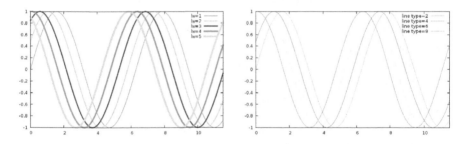

Figure 5.5: Different line widths   Figure 5.6: Different line types

```
    set arrow [<tag>][from <sx>,<sy>] [to <ex>,<ey>]
# To force auto scaling of an axis
    set autoscale [<axis>]
# To display border
    set [no]border
# tick mark label format specification
    set format [<axis>]["format-string"]
# To draw a grid at tic marks
    set [no]grid
# To enable/disable key (legend) of curves in plot
    set [no]key <x>,<y>,<z>
# To turn on time/date stamp
    set [no]time
```

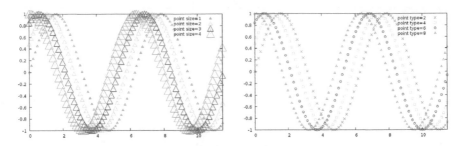

Figure 5.7: Different point sizes   Figure 5.8: Different point types

```
# To set centered plot title
    set title "title-text" <xoff>,<yoff>
# To control graphics device
    set terminal <device>
# To change direction of ticks
    set tics <direction>
# To adjust relative height of vertical axis
    set ticslevel <level>
# To set x-axis label
    set xlabel "<label>" <xoff>,<yoff>
# To set y-axis label
    set ylabel "<label>" <xoff>,<yoff>
# To set horizontal range
    set xrange [<xmin>:<xmax>]
# To set vertical range
    set yrange [<ymin>:<ymax>]
# To change horizontal ticks
    set xtics <start>,<incr>,<end>, "<label>" <pos>
# To change vertical ticks
    set ytics <start>,<incr>,<end>, "<label>" <pos>
...
```
Some of the set commands are given in Listing 5.32.

5.11.2 Histograms

Histograms are also known as bar charts. These can also be plotted using Gnuplot. Examine Listing 5.33.

Listing 5.33: Histogram plot

```
1  # This program plots a clustered bar chart using
2  # data from the file scheduling.dat;
3  # File name: histo.p;
4  # run: load 'histo.p'
5  set grid;
```

```
6  set style data histograms;
7  set style histogram cluster;
8  set style fill pattern;
9  set key left;
10 plot 'scheduling.dat' using 2:xtic(1) t "Earliest Deadline First
      ", \
11 'scheduling.dat' u 3 t "Min Util First", "" u 4 t "Max Util
      First";
```

The col_1 data of the data file (scheduling.dat) is taken to be the values on the x-axis. The values in col_2 to col_4 form three bars in a cluster. Each row of the data file is treated as one cluster, and the number of bars in a cluster is the number of columns considered in the plot command. The plot is shown in Figure 5.9.

5.11.3 Multiplot

Some times it may be required to draw multiple graphs in one plot for which the multiplot facilility of Gnuplot may be used. Using multiplot, more than one subplot may be plotted in one frame (like subplot in MATLAB®). The following steps need to be followed to plot a multiplot.

1) set the multiplot environment

2) define the size of a subplot

3) define the origin of a subplot (the left bottom corner of the frame is 0,0)

4) plot the data/function

5) repeat step 2 to step 4 for all subplots

6) unset the multiplot environment

Let us write a script to plot multiple graphs using a multiplot. In the example, two subplots are drawn in vertically separated frames. Refer to Listing 5.34.

Listing 5.34: Two plots in a multiplot

```
1  # --- multiplot.p ---
2  # Script produces multiple plots in a single picture
3  set xrange [-pi:pi]
4  set multiplot;    # get into multiplot mode
5  set size 0.5,1;
6  set origin 0.0,0.0;
7  plot sin(x);
8  set size 0.5,1;
9  set origin 0.5,0.0;
10 plot cos(x)/x
11 unset multiplot  # exit multiplot mode set
```

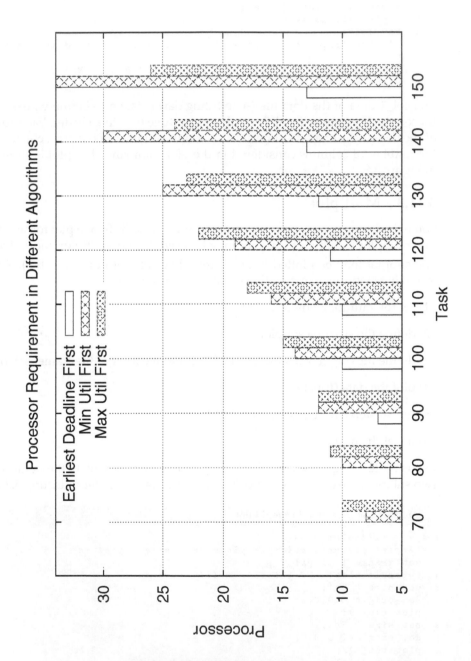

Figure 5.9: Clustered histograms

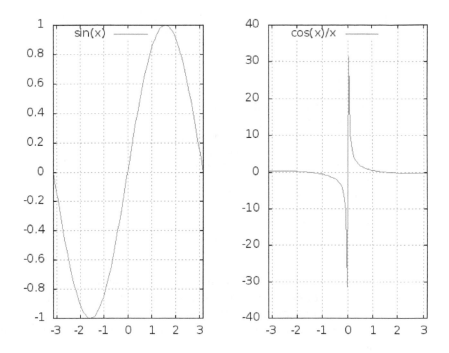

Figure 5.10: Multiplot separated vertically

As can be observed from the script, the multiplot environment is created in line 2, and the sizes of both subplots are defined as 0.5,1, i.e., a half portion of the x-axis and a full portion of the y-axis. The origins of the subplots are defined in line 4 and line 7, respectively. For the first plot, it is (0,0), i.e., first subplot starts from the left bottom corner and the origin for the second subplot is (0.5,0.0), which is half way on the x-axis. The resultant plot is shown in Figure 5.10.

5.11.4   Saving plots into files

Finally, it may be required to save the plot in an image or postscript file for further use, such as documentation, archiving, etc. Gnuplot supports a variety of output formats. To select a particular output, an appropriate terminal type must chosen. Each terminal has its additional modifying options for selecting fonts, font size, color, etc. To save a plot
1. First set the proper terminal along with configurations.
2. Then set the output file name.
We will discuss two terminals that are normally used.

1. Postscript Terminal
   Different options that may be set in the postscript driver are given as

follows.
Syntax:

```
set terminal postscript {default}
set terminal postscript
    {landscape | portrait | eps}
    {enhanced | noenhanced}
    {fontfile [add | delete] "<filename>"
    | nofontfiles} {{no}adobeglyphnames}
    {level1 | leveldefault}
    {color | colour | monochrome}
    {solid | dashed}
    {dashlength | dl <DL>}
    {linewidth | lw <LW>}
    {blacktext | colortext | colourtext}
    {{font} "fontname{,fontsize}"
            {<fontsize>}}
```

`landscape` and `portrait`: Defines the plot orientation.
`eps mode` generates output in Encapsulated PostScript format (EPS),
which is an extension to regular PostScript format.
`enhanced`: Features such as *subscripts, superscripts and mixed fonts* are
enabled.

`<fontname>` defines the name of any PostScript font, and `<fontsize>`
defines font size. Along with standard postscript fonts, another font
called "Symbol-Oblique" is used and is useful for mathematics.

`'default'` is used to set default values to different options such as
`landscape, monochrome, dashed, dl 1.0, lw 1.0, defaultplex,`
`noenhanced,` `"Helvetica"` and `14pt`.

Similarly, the default size of a PostScript plot is 10×7 square inches.
The options `color` or `monochrome` may be used to enable either *color* or
black and white drawing elements. The `'solid'` option is used to draw
solid-lined plots. The option `'dash-length (dl)'` is used to scale the
length of the dashed-line segments by `<dl>`, where dl is a floating-point
number with value greater than zero. Similarly, the option `'linewidth`
`(lw)'` is used to scale line widths by `<lw>`.

Example 1: `set terminal postscript default`

2. PNG Terminal
   Syntax: `set terminal png`
   For different options, the reader is referred to the Gnuplot user's
   manual.

Example 2: `set term png font arial 11`

This generates a plot in 640×480 pixels. The color of the background is specified in the form xrrggbb, where x remains as a character 'x' and the remaining charters are replaced with corresponding two-digit hexadecimal values for red (rr), green (gg), and blue (bb) components, i.e., a string 'x00ff00' represents green background color.

Example 3: `set terminal png medium size 640,480 x00ff00`

To have a nontransparent white background, use the string as 'xffffff'.

    `set terminal png font arial 14 size 800,600`

The above command finds a scalable font 'arial' and sets the font size to 14 pt.

    `set terminal png transparent truecolor enhanced`

The above command uses 24 bits of color information per pixel with a transparent background. It also uses the enhanced text mode that controls the layout of strings to be printed.

Now let us consider a complete example for plotting and saving the plot in a file.

Listing 5.35: Gnuplot script to save a plot

```
1  # This program plots a clustered Histogram and saves to a
       postscript file;
2  # File name: histo1.p;
3  set terminal postscript;
4  set output 'histo.ps';
5  set grid;
6  set style data histograms;
7  set style histogram cluster;
8  set style fill pattern;
9  set key left;
10 plot 'scheduling.dat' using 2:xtic(1) t "Earliest Deadline
       First", \
11 "" u 3 t "Min Util First", "" u 4 t "Max Util First";
```

Line 3 of Listing 5.35 sets the terminal to postsript, and line 4 provides the file name in which the plot would be saved. When executed, this script generates and saves the plot to a file histo.ps, which is a ps file and can be opened in any postscript viewer or pdf viewer. Similarly, a plot can be saved in other formats like png, etc.

Test/Viva Questions

    a) What are the different data sources which can be used to draw a plot in Gnuplot?

    b) What is the difference between a Gnuplot script file and a data file?

    c) What is the command to execute a script file in Gnuplot?

    d) What is the meaning of "customizing" a plot?

e) Distinguish between *points* and *linepoints* styles.

f) A graph is plotted using the options `w p ps 3`. Explain the meaning.

g) A graph is plotted using the options `w l lt 6`. Explain the meaning.

h) What is the command to put the legend at a particular coordinate?

i) What is the difference between a histogram and a line graph?

j) Differentiate between multiple plots and multiplot.

Programming Assignments

Consider a data file that contains a year of student strength and the number of students who passed in that year in an institution for 10 years. Using this, draw the following plots. (Fill the file with arbitrary data values.)

1) Draw a line graph showing year on the x-axis and students present on the y-axis.

2) Draw a graph showing the year on the x-axis and passed students on the y-axis using large points.

3) Draw a histogram to show the number of students available in a year and the number of students who passed.

4) Draw a points graph that contains two lines, one for students available and another for pass percentage in the same graph. Provide a proper legend, plot title, and axis title and also show grid lines.

5) Draw a multiplot that contains two subplots. One subplot is a line graph showing students available and the other one shows the number of students who passed.

Bibliography

[1] Tcl Tutorial:
https://www.tcl.tk/man/tcl8.5/tutorial/tcltutorial.html

[2] The GNU Awk User's Guide:
http://www.gnu.org/software/gawk/manual/gawk.html

[3] Official Gnuplot documentation:
http://www.gnuplot.info/documentation.html

CHAPTER 6

WIRED NETWORK SIMULATION

6.1 Introduction

The architecture of NS2 and certain preliminary concepts were discussed in Chapter 4 and Chapter 5. In this chapter, we discuss simulation of a wired network. Computer communication networks may be classified into two types: wired and wireless. In wired networks, the nodes are connected via cables. On the other hand, in wireless networks the nodes are connected with air as the medium. In the first case a topology of the network is usually defined, whereas in the second case nodes can move and the topology changes dynamically and hence no static topology can be defined. If a node comes within the communication range of another node, then direct communication between the two nodes is possible. In this chapter we discuss simulation of wired networks, and wireless network simulation is discussed in Chapter 7. We start with a few simple examples and gradually develop programs of increasing sophistication with the objective of being able to simulate real-life wired network scenarios for various kinds of networks.

6.2 Step-by-Step Wired Network Simulation

Complete network simulation using NS2 involves many steps. To be able to satisfactorily simulate a network, a thorough understanding of these steps is necessary. A minimal step-by-step process to simulate any wired network is provided in the form of a block diagram in Figure 6.1. The different blocks are briefly discussed below. Each block represents a step, and the details of these steps are discussed in subsequent sections. All simulation scripts are written using Tcl.

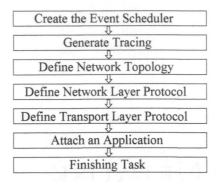

Figure 6.1: Block diagram for wired network simulation in NS2

Step I: Creating the Event Scheduler

Creating the event scheduler is the *first program statement* in any NS2 program. This scheduler queues any event that is generated by the simulation. The scheduler is an object of the simulator class. The following Tcl statement is used to create the event scheduler.

```
set ns [new Simulator]
```

Step II: Tracing

This step is essential if you need to record the events that are occurring during the simulation experiments in a specific format in a plain-text file. These files are treated as the output of any simulation program execution. Two types of traces are available. One is used by the programmer to analyze the simulation results for various network parameters (like throughput, delay, etc.) and is called the *Packet Trace/Event Trace/NS trace*. The other one is used by the network animator (NAM) module to create visualization for the simulation and is known as the *NAM Trace*. The following Tcl syntax may be used to generate traces.
Syntax: Packet Trace

```
set ptf [open <pktTrcFileName> w]

$ns trace-all $ptf
```
Syntax: NAM Trace

```
set ntf [open <namTrcFileName> w]

$ns namtrace-all $ntf
```
The file names given in angular brackets are user defined, i.e., the user has to provide the file name. Packet-trace is discussed in more detail in Section 6.7.

Step III Creating Network Topology

In this step the network topology is created. Different kinds of network topologies can be defined as per the user requirement. To realize a required topology, a set of nodes is first created, and the links between the nodes are defined as per the requirement. The syntax to create the nodes and links is provided below. Also, some complete programs are given to demonstrate the creation of different topologies. These programs may be executed to visualize the topologies in the NAM window.

Syntax: Creating nodes

```
set <nodeVar> [$ns node]
```

Syntax: Creating a link between two nodes

```
$ns <link-type>  <node1>  <node2>  <link-BW>  <delay>
<queue-type>
```

Link parameters:

- The *link-type* can be simplex or duplex.

- *node1* and *node2* represents the nodes between which the link needs to be established.

- *link-bw* represents the bandwidth of the link normally provided in Mbps.

- Delay is the propagation delay in ms.

- Each link is associated with an interface queue in the MAC sublayer. These queues can be of various types and are discussed in subsequent sections. The simplest *queue type* is drop tail, which is essentially a FIFO queue and when the queue is full any arriving packet is dropped.

Let's now write a program to create two nodes and a link between them. The code is given in Listing 6.1.

Listing 6.1: A Two Node Network

```
1  set ns [new Simulator]; #create Simulator Object
2
3  set nf [open twoNode.nam w];#create NAM trace file
4  $ns namtrace-all $nf; # write into nam file
5
6  set n0 [$ns node]; #create node n0
7  set n1 [$ns node]; #create node n1
8
9  #create a duplex link between them
10 $ns duplex-link $n0 $n1 10Mb 10ms DropTail
11
12 $ns run; # run the simulation
```

Write and save the program in a file with extension .tcl (*twoNode.tcl*). Now run the program using the following command in the command line/shell prompt.

$ *ns twoNode.tcl*

This command will execute the Tcl script and produce the NAM traces

Figure 6.2: A two-node network with a *point-to-point* link.

that are stored in a file named *twoNode.nam*, as given in line 3 of Listing 6.1. To see the NAM visualization, use the following command.
$ *nam twoNode.nam* &
This results in the creation of the topology shown in Figure 6.2. In this program only the NAM trace is used, and no packet trace is used, as no packet transmission is made between these nodes. The packet transmission mechanism will be discussed later in this chapter.

Let us now write another program for a four-node mesh network, as shown in Listing 6.2.

Listing 6.2: Four-node mesh network

```
1  #Four node Mesh Topology
2  set ns [new Simulator]
3
4  set nf [open fourNode.nam w]
5  $ns namtrace-all $nf
6
7  set n0 [$ns node]
8  set n1 [$ns node]
9  set n2 [$ns node]
10 set n3 [$ns node]
11
12 #create link between each node
13 $ns duplex-link $n0 $n1 10Mb 10ms DropTail
14 $ns duplex-link $n0 $n2 10Mb 10ms DropTail
15 $ns duplex-link $n0 $n3 10Mb 10ms DropTail
16 $ns duplex-link $n1 $n2 10Mb 10ms DropTail
17 $ns duplex-link $n1 $n3 10Mb 10ms DropTail
18 $ns duplex-link $n2 $n3 10Mb 10ms DropTail
19
20 $ns run
```

When code Listing 6.2 is executed using execution commands (as in the previous program), it will produce the topology provided in Figure 6.3

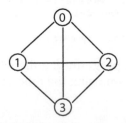

Figure 6.3: Four-node mesh topology

with four nodes and six links among them. If the actual layout does not resemble the figure provided, then click on the relayout button of the NAM window provided in the right bottom corner to get the figure in the proper layout (it may sometimes require multiple clicks).

As can be seen from Listing 6.2, to create multiple nodes and links lines 7 to 10 and lines 13 to 18 are repeated redundantly. The same program may be rewritten using loops and arrays, as shown in Listing 6.3.

Listing 6.3: Four-node mesh network using loops

```
1  #Four node Mesh Topology
2  set ns [new Simulator]
3
4  set nf [open fourNode.nam w]
5  $ns namtrace-all $nf
6  #create nodes
7  for {set i 0} {$i < 4} {incr i} {
8      set n($i) [$ns node]
9  }
10 #create links
11 for  {set i 0} {$i < 3} {incr i} {
12    for {set j [expr $i+1]} {$j < 4} {incr j} {
13            $ns duplex-link $n($i) $n($j) 10Mb 10ms DropTail
14      }
15 }
16
17 $ns run
```

Step IV: Simulating Network Layer Protocol

The network layer activity definition is optional in NS2. That is, if the user does not define any protocol for the network layer, then the simulator chooses a default routing protocol. We defer discussion of the definition of the network layer to subsequent sections.

Step V: Attaching Transport Protocol

Transport agents are created in pairs, as default options of NS2 support half-duplex connections in the transport layer. That is, one source agent and one sink agent need to be created for any connection. Many transport layer protocols are available in NS2, but here we discuss the UDP agent, the simplest one, and other agents are discussed later in this chapter.
Syntax: Create source agent

  set <**udpSrcAgent**> **[new Agent/UDP]**
Syntax: Attach the agent to a specific node (sender)

  $ns attach-agent <**node1**> <**udpSrcAgent**>
Syntax: Create sink agent

  set <**udpRecvAgent**> **[new Agent/Null]**
Syntax: Attach the sink agent to another node (receiver)

```
$ns attach-agent <node2> <udpRecvAgent>
```
Syntax: Connect two agents

```
$ns connect <udpSrcAgent> <udpRecvAgent>
```

Step VI: Application

Now we need to attach an application that generates packets for transmission through the connections (traffic generation). Out of many traffic types available in NS2, we choose constant bit rate traffic for this example. Other traffic types will be discussed later in this chapter.
Syntax: Create and attach application traffic

```
set <cbrTraffic> [new Application/Traffic/CBR]
```

```
<cbrTraffic> attach-agent <udpSrcAgent>
```
Syntax: start and stop data transmission

```
$ns at <time in sec> "<cbrTraffic> start"
```

```
$ns at <time in sec> "<cbrTraffic> stop"
```

Step VII Finishing Touch

The time period for which the simulation should run needs to be mentioned.
Syntax: Simulation time

```
$ns at <time in sec> "finish"
```
'finish' is a user-defined procedure, designed to perform some routine tasks at the end of the simulation like closing trace files, executing NAM visualization, etc.

Finally, run the simulation. Syntax: run the simulation

```
$ns run
```
This command should always be the *'last line'* of each simulation.

The complete program is given in Listing 6.4. Write and save the program in a file with extension .tcl and then execute the program from the command prompt (ns myFirstNSProgram.tcl). If everything goes fine (no syntax error), then it will execute the simulation for the defined duration of time and automatically open the NAM visualization. To observe the visualization of data transmission, the play button of the NAM window may be clicked.

Listing 6.4: Complete program for two-node network simulation

```
1  set ns [new Simulator]
2  set nf [open twoNode.nam w]
3  $ns namtrace-all $nf
4
5  set n0 [$ns node]
6  set n1 [$ns node]
```

```
7
8   $ns duplex-link $n0 $n1 100Mb 5ms DropTail
9
10  set udp [new Agent/UDP]
11  $ns attach-agent $n0 $udp
12  set null [new Agent/Null]
13  $ns attach-agent $n1 $null
14  $ns connect $udp $null
15  set cbr [new Application/Traffic/CBR]
16  $cbr attach-agent $udp
17  $ns at 1.0 "$cbr start"
18  $ns at 3.0 "$cbr stop"
19  $ns at 3.1 "finish"
20
21  proc finish { } {
22    global ns nf
23    $ns flush-trace
24    close $nf
25    exec nam twoNode.nam &
26    exit 0
27  }
28
29  $ns run
```

When the program in Listing 6.4 is executed, it produces a visualization in NAM due to line 25 of the program. The animation will show packet flow from node 0 to node 1.

6.3   Visualization Using NAM

NAM visualizion of Listing 6.4 shows the default shape and default color of different objects. That is, nodes are circular, and the color of nodes, links, and packets is black. However, using certain Tcl commands, the visualization can be made more colorful and meaningful. Some of these commands are discussed below.

- **Nodes**

    - *Color:* coloring a node

        $node color blue ; #creates a blue color node
          Other colors are *red, green, chocolate,* etc.

    - *Shape:* To draw nodes of different shapes

        $node shape box ; #creates a square shaped node
          Other shapes are *circle, box, hexagon*

        $n0 add-mark m0 blue box ; # creates a concentric circle over node n0

        $ns at 2.0 "$n0 delete-mark m0" ; # deletes the mark at time 2 sec.

    – *Label:* To provide a label to a node

         `$n0 label Router` ; # labels n0 as a router

- **Links**

    – *Color:* Coloring a link

    # To make a green color link from n0 to n1.

    `$ns duplex-link-op $n0 $n1 color "green"`

    – *Label:* To provide a label to a link

    # A link is labeled as 'point-to-point'

    `$ns duplex-link-op $n0 $n1 label "point-to-point"`

    – *Link Orientation:*

    # To draw a horizonatal link from n0 to n1

    `$ns duplex-link-op $n(0) $n(1) orient right`

    `$ns duplex-link-op $n(1) $n(2) orient left`

    # Other Orientations

    `$ns duplex-link-op $n(0) $n(2) orient up`

    `$ns duplex-link-op $n(1) $n(3) orient down`

    `$ns duplex-link-op $n(0) $n(3) orient right-up`

    `$ns duplex-link-op $n(2) $n(1) orient left-down`

    `$ns duplex-link-op $n(5) $n(6) orient 60deg`

- **Packets**

    – *Color:*

    Step I: (map the colors to integers)

        `$ns color 40 red`

        `$ns color 41 blue`

        `$ns color 20 green`

    Step II: (flow id association)

        `$tcp0 set fid_ 40` ; # traffic-1 produces red packets

        `$tcp1 set fid_ 41` ; # traffic-2 produces blue packets

        `$udp0 set class_ 20` ; # traffic-3 produces green packets

- **Miscellaneous**

    – *Annotation:*

    # To add textual explanation to the simulation

    `$ns at 3.5 "$ns trace-annotate \"packet drop\""`

    – *Animation Rate:*

    # To customize rate of animation

    `$ns at 0.0 "$ns set-animation-rate 0.1ms"`

Other commands for NAM visualization are discussed in their respective sections. Listing 6.5 incorporates some of these visualization effects.

Listing 6.5: Colorful NAM visualization

```
1   set ns [new Simulator]
2
3   $ns color 0 blue
4   $ns color 1 red
5   $ns color 2 green
6
7   set n0 [$ns node]
8   $n0 color purple ;# Coloring nodes
9
10  set n1 [$ns node]
11  $n1 color purple
12
13  set n2 [$ns node]
14  $n2 shape box
15  $n2 color red
16  $ns at 0.0 "$n2 label Router" ;# Labeling node
17  set n3 [$ns node]
18
19  $ns at 1.0 "$n0 add-mark m0 blue box" ;# Concentric circle
20  $ns at 2.0 "$n0 delete-mark m0"
21
22  # Annotations
23  $ns at 1.0 "$ns trace-annotate \"simulation starts now\""
24  $ns at 0.0 "$ns set-animation-rate 500us"
25
26  set nf [open out.nam w]
27  $ns namtrace-all $nf
28
29  $ns duplex-link $n0 $n2 5Mb 2ms DropTail
30  $ns duplex-link $n1 $n2 5Mb 2ms DropTail
31  $ns duplex-link $n2 $n3 1.5Mb 10ms SFQ
32
33   # Link Orientations
34  $ns duplex-link-op $n0 $n2 orient right-up
35  $ns duplex-link-op $n1 $n2 orient right-down
36  $ns duplex-link-op $n2 $n3 orient right
37  $ns duplex-link-op $n2 $n3 color "green"
38  $ns duplex-link-op $n2 $n3 label "Bottleneck"
39  $ns duplex-link-op $n2 $n3 queuePos 0.5 ;# visualize Queue
40
41  set udp0 [new Agent/UDP]
42  $ns attach-agent $n0  $udp0
43  set cbr0 [new Application/Traffic/CBR]
44  $cbr0 attach-agent $udp0
45
46  set udp1 [new Agent/UDP]
47  $ns attach-agent $n3 $udp1
48  $udp1 set class_ 2 ;# Packet Color green
49  set cbr1 [new Application/Traffic/CBR]
50  $cbr1 attach-agent $udp1
51
```

```
52   set null0 [new Agent/Null]
53   $ns attach-agent $n3 $null0
54   set null1 [new Agent/Null]
55   $ns attach-agent $n1 $null1
56   $ns connect $udp0 $null0
57   $ns connect $udp1 $null1
58
59   $ns at 0.5 "$cbr0 start"
60   $ns at 1.5 "$cbr1 start"
61   $ns at 2.3 "$cbr0 stop"
62   $ns at 1.9 "$cbr1 stop"
63
64   set tcp [new Agent/TCP]
65   $tcp set class_ 1
66   set sink [new Agent/TCPSink]
67   $ns attach-agent $n1 $tcp
68   $ns attach-agent $n3 $sink
69   $ns connect $tcp $sink
70
71   set ftp [new Application/FTP]
72   $ftp attach-agent $tcp
73   $ns at 1.0 "$ftp start"
74   $ns at 2.0 "$ns detach-agent $n0 $tcp
75   $ns detach-agent $n3 $sink"
76
77   $ns at 3.0 "finish"
78
79   proc finish {} {
80     global ns f nf
81     $ns flush-trace
82     close $nf
83     exec nam out.nam &
84     exit 0
85   }
86
87   $ns run
```

Test/Viva Voce Questions

    a) What are the different steps (in sequence) required for the simulation of a network?

    b) What is the major work of an event scheduler?

    c) What are the different kinds of traces available in NS2?

    d) Explain the syntax to create a link.

    e) What are the different shapes available to visualize a node?

    f) How does one provide a packet color?

    g) What is animation rate?

    h) What is the significance of a queue in a link?

    i) What is the command for link orientation?

    j) What is the use of a traffic generator?

Programming Assignments

1) Write a Tcl program to create a ring network of six nodes with red color links between them.

2) Write a Tcl program to create a star network with 15 leaf nodes of red color and a box-shaped central node of blue color with the label "switch."

3) Write a Tcl program to create a 10-node mesh network with 5 green nodes and 5 blue nodes.

4) Write a Tcl program to create a four-node network, where node n0 and n1 are connected to node n2. Node n2 is connected to n3. All links are of 10 Mbps bandwidth. Attach a UDP agent and CBR traffic to both n0 and n1 and attach both the null agents to n3. Assign different colors to different traffic and run the simulation script for 50 seconds.

5) Write a Tcl program to create a 10-node ring network. Multiple nodes and links are to be created using loops, and all nodes/links are to be stored in a different array. Node n1 sends CBR traffic to node n3 and another CBR traffic is sent from node n0 to node n8 for 20 sec and 30 sec, respectively.

6) Write a Tcl program to create two star networks with six and five leaf nodes, respectively; label the networks. The central nodes of both the networks are connected. Attach traffic as follows.

   i) A traffic within network_1.

   ii) Another traffic within network_2.

   iii) A traffic from network_1 and network_2.

7) Write a Tcl program to connect a ring network of five nodes with a star network of six nodes. Send traffic between the two networks.

8) Write a Tcl program to create a hypercube network of eight nodes. Attach different types of traffic to different nodes.

9) Write a Tcl program to create three star networks such that their central nodes are connected in a ring structure. Now attach one traffic within a network and a traffic across the three star networks.

10) Write a Tcl program to create four star networks and interconnect all these networks as another star network. Attach some traffic between different nodes and networks.

6.4   Link Layer — Links and Queueing

Links can be *point-to-point, multipoint, broadcast* links, and so on. In this section the use of first two types of links is discussed.

6.4.1   Point-to-point links

   `$ns simplex-link <n1>   <n2>   <BW>   <Delay>   <Qtype>`
`<Args>`

   The above command establishes a simplex (unidirectional) link between node n1 and node n2 with the specified bandwidth (BW) and propagation delay. The <Qtype> parameter defines the type of queue buffer to be used by the link, and, according to the queue type mentioned, different arguments may be passed through the <args> parameter. This link can send data from n1 to n2; the reverse communication is not possible.

   However, to perform duplex communication between nodes, the following command may be used.

   `$ns duplex-link <n0> <n1> <BW> <Delay> <Qtype>`
This command establishes a bi-directional link between node n0 and node n1, with specified bandwidth and propagation delay. As usual, the <Qtype> parameter defines the type of queue to be used by the link.

   The following command is used to set different duplex-link attributes, such as physical orientation of the links, color, label, or queue position in NAM visualization.

   `$ns duplex-link-op <n1>   <n2>   <op>   <args>`

   `$ns link-lossmodel <lossobj>   <from>   <to>`
The above command introduces losses in to the link between <n1> node and <n2> node.

   `$ns lossmodel <lossobj>   <from>   <to>`
Above command is used to insert a loss module in regular links.

   The queue limit may be specified as

   `$ns queue-limit $n1 $n2 <number of packets>`

   Queues are used to hold or drop the packets. Packet scheduling is done to decide whether a packet is to be inserted in the queue for further processing or dropped. Queueing discipline is nothing but the management of the queue buffer to regulate a queue in a particular way. Many queueing disciplines are supported; some of them are discussed below.

   1) **DropTail:** This implements a simple FIFO queue. In this queueing mechanism, each packet is treated equally, and, when the queue is filled to its maximum capacity, the newly incoming packets are dropped until the queue has sufficient space to accommodate incoming traffic.

   2) **FQ:** This implements fair queueing, in which multiple flows are allowed to share the link capacity in a fair way. Routers maintain separate queues for each flow. These queues are served in a round robin fashion such that, if a flow sends more packets, then its queue becomes full quickly and the chance of packet drops for that flow increases whereas other flow may not drop any packet, thus providing a fair share of

bandwidth.
configuration parameter: *secsPerByte_*

3) **SFQ:** This implements stochastic fair queueing. Unlike the fair queueing technique, where each flow requires a separate queue (which may not be practicable), SFQ maintains a limited number of queues, and a hashing algorithm is used to map the traffic to one of the available queues.
Configuration parameters: *maxqueue_ bucket_*

4) **DRR:** This is deficit round robin scheduling queueing and is implemented as a modified weighted round robin scheduling mechanism. It can handle packets of different sizes. The flow is assigned to a queue using a hashing SFQ. Each queue is assigned a time slot and can send a packet of a size that can fit in the available time slot. If not, then the idle time slot is added to this queue's deficit, and the packet can be sent in the next round.

5) **RED:** Random early detection (RED) is a congestion avoidance queueing mechanism. It operates on the average queue size and drops/marks packets on the basis of statistics information. If the buffer is empty, all incoming packets are acknowledged. As the queue size increases, the probability of dropping a packet increases. When the buffer is full, the probability of dropping a packet becomes 1 and all incoming packets are dropped.

6) **CBQ:** In class based queueing (CBQ) the traffic is first classified in to different groups according to different parameters such as priority of the traffic, interface from which it is received, or originating program, etc. Then this queueing allows traffic to share bandwidth equally.

Listing 6.6 demonstrates two CBR traffic flows in a star network. One traffic is colored red and another is colored blue. The queue for the traffic, is shown in Figure 6.4. When the number of packets exceed the queue limit the packets are dropped from the tail of the queue.

Listing 6.6: Star network with two CBR traffics

```
1  #Creating event Scheduler
2  set ns [new Simulator]
3
4  $ns color 1 red
5  $ns color 2 blue
6
7  #Creating Nodes
8  set n0 [$ns node]
9  set n1 [$ns node]
10 set n2 [$ns node]
11 set n3 [$ns node]
12 set n4 [$ns node]
```

```
13
14   $ns trace-all [open starLoad.tr w]
15   set nf [open starLoad.nam w]
16   $ns namtrace-all $nf
17
18   #creating link between nodes with DropTail Queue
19   $ns duplex-link $n4 $n1 900Kb 10ms DropTail
20   $ns duplex-link $n4 $n2 800Kb 10ms DropTail
21   $ns duplex-link $n4 $n3 1Mb 10ms DropTail
22   $ns duplex-link $n4 $n0 1Mb 10ms DropTail
23
24   #Link Orientation
25   $ns duplex-link-op $n4 $n1 orient left-down
26   $ns duplex-link-op $n4 $n2 orient right-up
27   $ns duplex-link-op $n4 $n3 orient right-down
28   $ns duplex-link-op $n4 $n0 orient left-up
29   $ns queue-limit $n4 $n1 10
30   $ns queue-limit $n4 $n2 20
31   $ns duplex-link-op $n4 $n1 queuePos 1.75
32   $ns duplex-link-op $n4 $n2 queuePos 1.75
33
34   ###UDP Agent between n0 and n2
35   set udp0 [new Agent/UDP]
36   set null0 [new Agent/Null]
37   $ns attach-agent $n0 $udp0
38
39   $ns attach-agent $n2 $null0
40   $ns connect $udp0 $null0
41   $udp0 set fid_ 1
42
43   ### CBR traffic between n0 and n2 (Red)
44   set cbr [new Application/Traffic/CBR]
45   $cbr set packetSize_ 700
46   $cbr set interval_ 0.005
47   $cbr attach-agent $udp0
48
49   ###UDP Agent between n3 and n1
50   set udp1 [new Agent/UDP]
51   set null1 [new Agent/Null]
52   $ns attach-agent $n3 $udp1
53   $ns attach-agent $n1 $null1
54   $ns connect $udp1 $null1
55   $udp1 set fid_ 2
56
57   ### CBR traffic between n3 and n1 (Blue)
58   set cbr1 [new Application/Traffic/CBR]
59   $cbr1 set packetSize_ 500
60   $cbr1 set interval_ 0.004
61   $cbr1 attach-agent $udp1
62
63   $ns at 0.1 "$cbr start"
64   $ns at 0.2 "$cbr1 start"
65   $ns at 2.35 "$cbr stop"
66   $ns at 2.8 "$cbr1 stop"
67   $ns at 3.0 "finish"
68
```

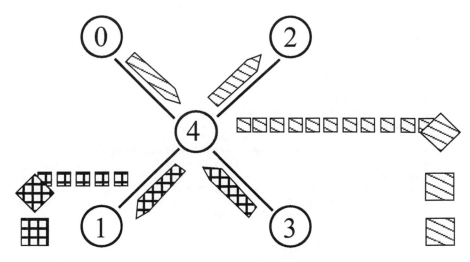

Figure 6.4: Star network with two CBR traffics and queue

```
69  proc finish {} {
70    global nf
71    close $nf
72    exec nam starLoad.nam &
73    exit 0
74  }
75
76  #Run Simulation
77  $ns run
```

In the topology created in Listing 6.7, two CBR traffics share the common link. The output of the program is given in Figure 6.5. Because of the drop tail queue defined, more blue (slanted line packets) packets are dropped than red packets (square marked packets). Now change line 22 of the program in Listing 6.7 as follows:

$ns duplex-link $n2 $n3 1Mb 10ms FQ

The queue now uses the fair queueing technique. The output is shown in Figure 6.6, where an almost equal number of packets is dropped from both instances of traffic. Similarly, the other queueing mechanisms like SFQ and RED may be used to experience the difference in the mechanisms.

Listing 6.7: Queue shared by multiple flows

```
1  #Creating event Scheduler
2  set ns [new Simulator]
3
4  $ns color 1 red
5  $ns color 2 blue
6
7  #Creating Nodes
8  set n0 [$ns node]
```

```
9   set n1 [$ns node]
10  set n2 [$ns node]
11  set n3 [$ns node]
12  set n4 [$ns node]
13  set n5 [$ns node]
14
15  #Network Animator
16  set nf [open starLoad.nam w]
17  $ns namtrace-all $nf
18
19  #creating link between nodes with DropTail Queue
20  $ns duplex-link $n0 $n2 1Mb 10ms DropTail
21  $ns duplex-link $n1 $n2 1Mb 10ms DropTail
22  $ns duplex-link $n2 $n3 1Mb 10ms DropTail
23  $ns duplex-link $n3 $n4 1Mb 10ms DropTail
24  $ns duplex-link $n3 $n5 1Mb 10ms DropTail
25
26  #link Orientation
27  $ns duplex-link-op $n0 $n2 orient right-down
28  $ns duplex-link-op $n1 $n2 orient right-up
29  $ns duplex-link-op $n2 $n3 orient right
30  $ns duplex-link-op $n3 $n4 orient right-up
31  $ns duplex-link-op $n3 $n5 orient right-down
32
33  #creating link between nodes with FQ Queue
34  $ns duplex-link-op $n2 $n3 queuePos 0.5
35
36  ###UDP Agent between n0 and n5
37  set udp0 [new Agent/UDP]
38  set null0 [new Agent/Null]
39  $ns attach-agent $n0 $udp0
40  $ns attach-agent $n5 $null0
41  $ns connect $udp0 $null0
42
43  $udp0 set fid_ 1
44  ### CBR traffic between n0 and n2 (Red)
45  set cbr [new Application/Traffic/CBR]
46  $cbr set packetSize_ 500
47  $cbr set interval_ 0.005
48  $cbr attach-agent $udp0
49
50  ###UDP Agent between n1 and n6
51  set udp1 [new Agent/UDP]
52  set null1 [new Agent/Null]
53  $ns attach-agent $n1 $udp1
54  $ns attach-agent $n4 $null1
55  $ns connect $udp1 $null1
56
57  $udp1 set fid_ 2
58  ### CBR traffic between n3 and n1 (Blue)
59  set cbr1 [new Application/Traffic/CBR]
60  $cbr1 set packetSize_ 500
61  $cbr1 set interval_ 0.005
62  $cbr1 attach-agent $udp1
63
64  $ns at 0.1 "$cbr start"
```

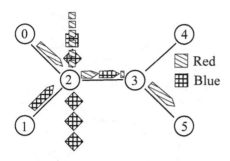

Figure 6.5: More blue packets dropped with droptail queue

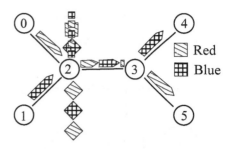

Figure 6.6: Almost an equal number of packets from both flows dropped with FQ

```
65  $ns at 0.1 "$cbr1 start"
66  $ns at 2.35 "$cbr stop"
67  $ns at 2.8 "$cbr1 stop"
68  $ns at 3.0 "finish"
69
70  proc finish {} {
71    global nf
72      close $nf
73      exec nam starLoad.nam &
74      exit 0
75  }
76
77  #Run Simulation
78  $ns run
```

6.4.2 Multipoint links

The behavior of a multipoint link is different from point-to-point links since it involves sharing of the link, due to which contention may occur when more than one node attempts to use the link at the same time. To simulate these properties, a special node called the LanNode is designed. The LanNode is a virtual node that represents the BUS of a LAN. The actual LAN is implemented in NS2 as given in Figure 6.7

Commands

   $ns make-lan <nodeList> <BW> <delay> <LL> <ifq> <MAC>
<channel>

or

   $ns newLan <nodeList> <BW> <delay> <args>

where

nodeList: A set of LAN member nodes within double quotes separated with a space

BW: LAN bandwidth in Mb or Kb

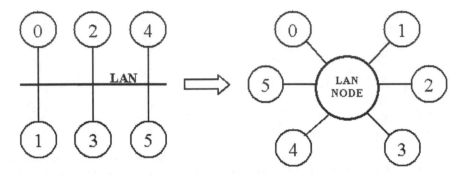

Figure 6.7: Actual LAN configuration (left), as implemented in NS2 (right)

delay: LAN delay in sec
args: Different argument types passed to the command are as follows
      LL: LL
      ifq: Queue/DropTail or any queueing strategy provided in Section 6.4.1
      MAC: MAC/802_3, MAC/Csma/Cd, MAC/Csma/Cd
      channel: Channel
Listing 6.8 demonstrates a LAN simulation.

Listing 6.8: Multipoint LAN simulation

```
1   #LAN simulation
2   set ns [new Simulator]
3
4   #define color for data flows
5   $ns color 1 Blue
6
7   set namfile [open lan.nam w]
8   $ns namtrace-all $namfile
9
10  proc  finish { } {
11    global ns namfile
12    $ns flush-trace
13    close $namfile
14    exec nam lan.nam &
15    exit 0
16  }
17
18  #create seven nodes
19  for {set i 0} {$i < 7} {incr i} {
20    set n$i [$ns node]
21  }
22
23  #create links between the node and Lan
24  $ns duplex-link $n5 $n6 0.3Mb 100ms DropTail
25  $ns duplex-link-op $n5 $n6 orient right
26
```

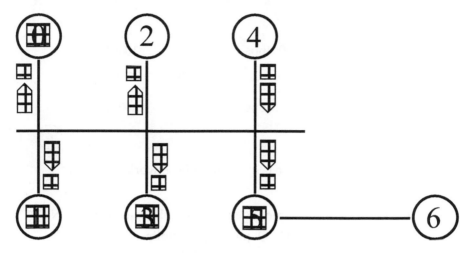

Figure 6.8: LAN simulation

```
27  #create Lan
28  set lan [$ns newLan "$n0 $n1 $n2 $n3 $n4 $n5" 0.5Mb 40ms LL
        Queue/DropTail MAC/Csma/Cd Channel]
29
30  #setup a UDP connection between n4--->n2
31  set udp [new Agent/UDP]
32  $ns attach-agent $n4 $udp
33  set null [new Agent/Null]
34  $ns attach-agent $n2 $null
35  $ns connect $udp $null
36  $udp set fid_ 1
37  set cbr [new Application/Traffic/CBR]
38  $cbr attach-agent $udp
39  $cbr set rate_ 50Kb
40
41  $ns at 0.1 "$cbr start"
42  $ns at 4.5 "$cbr stop"
43  $ns at 5.0 "finish"
44
45  $ns run
```

One snapshot of the execution of Listing 6.8 is given in Figure 6.8. Here n4 is sending CBR packets to n2. As seen from the figure, packets transmitted by n4 go to all nodes but are only accepted by n2, and other nodes drop the packets. This is because each LAN is considered to be one broadcasting domain. However, a unicast packet is received by the receiver only; other nodes discard the packets.

Test/Viva Voce Questions

    a) Differentiate between point-to-point and multipoint links.

    b) What is the syntax to create a link between two nodes?

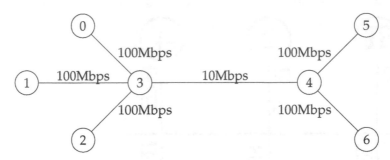

Figure 6.9: Seven-node network

c) What are the different queueing disciplines available in NS2?

d) Compare FQ and SFQ queueing disciplines.

e) How is a LAN programmed in NS2?

f) What are the different parameters needed in make-lan and new-lan commands?

g) What are different MAC protocols available in NS2?

h) What is the use of a LAN node in NS2?

i) How does packet transmission in a multipoint network differ from that in a point-to-point network?

j) What is a loss module, and how can it be attached to a link?

Programming Assignments

1) Design a network as shown in Figure 6.9, Now attach one traffic from 0 → 6, limit the queue size to 10 with (i) droptail queue (ii) fair queueing discplines. Observe, count, and comment on the difference in packet drops for different queueing disciplines.

2) Design a network as shown in Figure 6.9, Now attach two traffics from 0 → 6 and 2 → 5. Run the program with queue limit 15 and associate different colors with different traffic. Observe, count, and comment on the total number of packets dropped for each source when FQ and SFQ queueing disciplines are used.

3) Design a network as shown in Figure 6.9. Attach three traffics from 0 → 6, 2 → 5, and 1 → 4. Run the program with queue limit 20 and associate different colors with different traffic. Observe, count, and comment on the packets dropped for each source, when droptail, FQ, and SFQ queueing disciplines are used.

4) Create a 15-node LAN, with two sources and two destinations. Count the packets dropped/collided for each traffic.

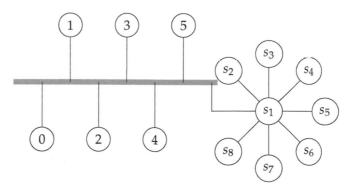

Figure 6.10: LAN connected to a star

5) Create a five-node LAN connected to a star network, as shown in Figure 6.10. Create one traffic within the LAN and another from the star to the LAN. Determine the number of packets dropped/-collided for each traffic.

6.5 Network Layer — Routing

In NS2 both unicast and multicast routings are supported. However, we restrict our discussion to unicast routing only. Readers are referred to the NS manual [1] for other routing technologies supported. A routing strategy is a mechanism to find/compute a route from a node to the destination node. The routing strategies available in NS2 are of three types, namely, static routing, dynamic routing, and manual routing. A static routing strategy calculates the routes between an arbitrary source and destination nodes once before simulation starts and remains unchanged throughout the simulation. In a dynamic strategy, the route is computed dynamically while executing the simulation. Therefore any change in topology (like link failure) is reflected in the routing table, and the routing changes accordingly. In manual routing, the user needs to specify the routes in the simulation script and that is used in the simulation. It cannot change the routing during the simulation as in static routing. Both static routing and session routing use Dijkstra's all-pairs shortest paths algorithm, whereas distance vector routing uses the distributed Bellman–Ford algorithm.

NS uses *static routing* implicitly if no routing commands are provided; this is called default routing.

$ns rtproto Session ;# uses *session routing*

$ns rtproto DV $n1 $n2 $n3 ;# uses *distance vector* routing on specified nodes

$ns rtproto Session $n1 $n2 ;# uses *link state* routing on given nodes

(a) *Static Routing Protocol:* This is the default route-finding technique in

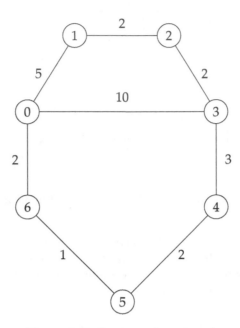

Figure 6.11: Seven node network

which Dijkstra's all-pairs shortest paths algorithm is used to compute the route between the source and the destination. The routing algorithm runs only once after the topology is formed, and the routing table formed is stored until the end of the simulation interval. Therefore it cannot accommodate any link failure due to topology changes while executing the simulation.

Let us consider a network as shown in Figure 6.11, where each link has a communication cost associated with it. Suppose that node 0 communicates with node 3; then there are three possible paths as follows.
path I: $0 \rightarrow 3$: 1 hop : cost = 10
path II: $0 \rightarrow 1 \rightarrow 2 \rightarrow 3$: 3 hops : cost = 9
path III: $0 \rightarrow 6 \rightarrow 5 \rightarrow 4 \rightarrow 4$: 4 hops : cost = 8
Out of these paths, path III is the lowest cost path and path I is the highest cost path. Therefore path III is chosen by the session routing protocol. Listing 6.9 give a program that uses static routing. A NAM screenshot of this simulation is presented in Figure 6.12; it has computed the shortest path from node 0 to node 3, that is, path III. But when the link $4 \rightarrow 5$ fails due to line 50 of Listing 6.9, the packets cannot be routed any further, as shown in Figure 6.13.

Listing 6.9: Static routing simulation

```
1 set ns [new Simulator]
```

```
2  $ns color 1 red
3
4  #creating nodes
5  for  {set i 0} {$i < 7} {incr i} {
6    set n($i) [$ns node]
7  }
8
9  #trace
10 set nf [open out.nam w]
11 $ns namtrace-all $nf
12
13 #creating links
14 for {set i 0} {$i < 7} {incr i} {
15   $ns duplex-link $n($i) $n([expr ($i+1)%7]) 1Mb 10ms
       DropTail
16 }
17 $ns duplex-link $n(0) $n(3) 1Mb 10ms DropTail
18
19 #routing protocol
20 #No routing is defined, That is, Static is used
21
22 #Create a UDP agent and attach it to node n(0) and sink to n
     (3)
23 set udp0 [new Agent/UDP]
24 $ns attach-agent $n(0) $udp0
25 set null0 [new Agent/Null]
26 $ns attach-agent $n(3) $null0
27 $udp0 set fid_ 1
28
29 # Create a CBR traffic source and attach it to udp0
30 set cbr0 [new Application/Traffic/CBR]
31 $cbr0 set packetSize_ 500
32 $cbr0 set interval_ 0.01
33 $cbr0 attach-agent $udp0
34 $ns connect $udp0 $null0
35
36 #Assign costs to links
37 $ns cost $n(0) $n(3) 10
38 $ns cost $n(0) $n(1) 5
39 $ns cost $n(1) $n(2) 2
40 $ns cost $n(2) $n(3) 2
41 $ns cost $n(3) $n(4) 3
42 $ns cost $n(4) $n(5) 2
43 $ns cost $n(5) $n(6) 1
44 $ns cost $n(6) $n(0) 2
45
46 $ns at 0.5 "$cbr0 start"
47 $ns at 4.5 "$cbr0 stop"
48
49 #link failure
50 $ns rtmodel-at 1.5 down $n(4) $n(5)
51 $ns rtmodel-at 2.5 up $n(4) $n(5)
52
53 $ns at 5.0 "finish"
54 proc finish {} {
55   global ns f nf
```

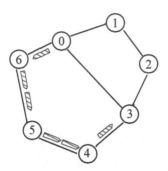

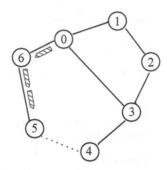

Figure 6.12: Static routing uses the shortest cost path

Figure 6.13: Static routing is unable to get new paths when the link 4 → 5 fails

```
56        $ns flush-trace
57        close $nf
58        puts "running nam..."
59        exec nam out.nam &
60        exit 0
61 }
62
63 # Run the simulation
64 $ns run
```

(b) *Session Routing Protocol:* This supports dynamic topology. The algorithm used is similar to the static routing algorithm. However, unlike in static routing, the route re-computation and recovery are done dynamically as and when the topology changes during the simulation period.

   Let us again consider Figure 6.11 and use session routing instead of static routing, where the routes are updated as and when the topology changes (link fails). The program code for the network in Figure 6.11 is provided in Listing 6.10.

Listing 6.10: Session routing simulation

```
1 set ns [new Simulator]
2
3 $ns color 1 red
4 $ns color 2 blue
5 $ns color 3 chocolate
6
7 #creating nodes
8 for {set i 0} {$i < 7} {incr i} {
9   set n($i) [$ns node]
10 }
11
12 set nf [open out.nam w]
13 $ns namtrace-all $nf
14
```

```
15  #creating links
16  for {set i 0} {$i < 7} {incr i} {
17    $ns duplex-link $n($i) $n([expr ($i+1)%7]) 1Mb 10ms DropTail
18  }
19  $ns duplex-link $n(0) $n(3) 1Mb 10ms DropTail
20  $ns duplex-link-op $n(0) $n(3) label ":10"
21  $ns duplex-link-op $n(0) $n(1) label ":5"
22  $ns duplex-link-op $n(1) $n(2) label ":2"
23  $ns duplex-link-op $n(2) $n(3) label ":2"
24  $ns duplex-link-op $n(3) $n(4) label ":3"
25  $ns duplex-link-op $n(4) $n(5) label ":2"
26  $ns duplex-link-op $n(5) $n(6) label ":1"
27  $ns duplex-link-op $n(6) $n(0) label ":2"
28
29  #link State Routing protocol
30  $ns rtproto Session
31
32  #Assign costs to links
33  $ns cost $n(0) $n(3) 10
34  $ns cost $n(0) $n(1) 5
35  $ns cost $n(1) $n(2) 2
36  $ns cost $n(2) $n(3) 2
37  $ns cost $n(3) $n(4) 3
38  $ns cost $n(4) $n(5) 2
39  $ns cost $n(5) $n(6) 1
40  $ns cost $n(6) $n(0) 2
41  set udp0 [new Agent/UDP]
42  $ns attach-agent $n(0) $udp0
43  set null0 [new Agent/Null]
44  $ns attach-agent $n(3) $null0
45  $udp0 set fid_ 1
46
47  set cbr0 [new Application/Traffic/CBR]
48  $cbr0 set packetSize_ 500
49  $cbr0 set interval_ 0.008
50  $cbr0 attach-agent $udp0
51  $ns connect $udp0 $null0
52
53  $ns at 0.5 "$cbr0 start"
54  $ns at 4.5 "$cbr0 stop"
55
56  #link failure
57  $ns rtmodel-at 1.5 down $n(3) $n(4)
58  $ns rtmodel-at 3.5 up $n(3) $n(4)
59  $ns rtmodel-at 2.0 down $n(1) $n(2)
60  $ns rtmodel-at 2.5 up $n(1) $n(2)
61
62  $ns at 1.5 "$udp0 set fid_ 2"
63  $ns at 2.0 "$udp0 set fid_ 3"
64  $ns at 2.5 "$udp0 set fid_ 2"
65  $ns at 3.5 "$udp0 set fid_ 1"
66
67  $ns at 5.0 "finish"
68  proc finish {} {
69    global ns nf
70    $ns flush-trace
```

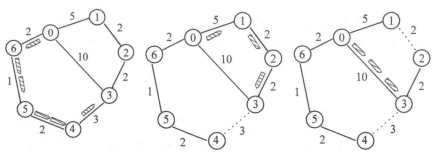

Figure 6.14: Session routing uses shortest cost path

Figure 6.15: Link n3 → n4 fails

Figure 6.16: Again the link from n1 → n2 fails

```
71    close $nf
72    puts "running nam..."
73    exec nam out.nam &
74    exit 0
75 }
76
77 $ns run
```

Screenshots of this simulation are provided in Figures 6.14, 6.15, and 6.16. As can be seen from these figures, initially it chooses path III and data transfer occurs in this path. After some time, the link 3 → 4 breaks due to the code in lines 57 to 60 of Listing 6.10. Now it chooses path II for communication. Again, after some time, the link from 1 → 2 goes down. Therefore only path I is chosen for transmission. Again, when the broken links come up, dynamically it changes the route according to the shortest path at that time.

(c) *Distance Vector Routing Protocol:* This protocol uses distance vector routing (or the distributed Bellman–Ford algorithm), which chooses the least hops path. It sends periodic route updates (default 2 sec) every advertised interval. Triggered updates are also sent either when a change in the topology occurs or when the node receives a route update. A *"split horizon with poisoned reverse"* mechanism is also implemented in which the node advertises the route with the hop metric as infinity. Each routing agent uses 32 for infinity to determine the validity of a route. Again, considering the network given in Figure 6.11, Listing 6.11 demonstrates the code to use for DV routing. The NAM screenshots are given in Figures 6.17, 6.18, and 6.19. It can be observed that the link cost has no impact on this routing but it finds the fewest number of hops. Therefore initially it chooses path I, and when the link 0 → 6 fails, it chooses the next longer path II and then path III.

Listing 6.11: DV routing simulation

```
1 set ns [new Simulator]
```

```
2
3  #creating nodes
4  for  {set i 0} {$i < 7} {incr i} {
5    set n($i) [$ns node]
6  }
7
8  set nf [open out.nam w]
9  $ns namtrace-all $nf
10
11 #creating links
12 for {set i 0} {$i < 7} {incr i} {
13   $ns duplex-link $n($i) $n([expr ($i+1)%7]) 1Mb 10ms
        DropTail
14 }
15 $ns duplex-link $n(0) $n(3) 1Mb 10ms DropTail
16
17 #Link State Routing protocol
18 $ns rtproto DV
19
20 set udp0 [new Agent/UDP]
21 $ns attach-agent $n(0) $udp0
22 set null0 [new Agent/Null]
23 $ns attach-agent $n(3) $null0
24 $udp0 set fid_ 1
25 set cbr0 [new Application/Traffic/CBR]
26 $cbr0 set packetSize_ 500
27 $cbr0 set interval_ 0.008
28 $cbr0 attach-agent $udp0
29 $ns connect $udp0 $null0
30
31 $ns at 0.5 "$cbr0 start"
32 $ns at 4.5 "$cbr0 stop"
33
34 #link failure
35 $ns rtmodel-at 1.5 down $n(3) $n(4)
36 $ns rtmodel-at 3.5 up $n(3) $n(4)
37 $ns rtmodel-at 2.0 down $n(1) $n(2)
38 $ns rtmodel-at 2.5 up $n(1) $n(2)
39
40 $ns at 5.0 "finish"
41 proc finish {} {
42   global ns nf
43   $ns flush-trace
44   close $nf
45   puts "running nam..."
46   exec nam out.nam &
47   exit 0
48 }
49
50 $ns run
```

(d) **Manual Routing:** Manual routing is a fixed routing scheme in which routes are manually set by the user instead of computing them by using an algorithm. Once the routes are set, they remains static and cannot be changed throughout the simulation period. This technique may be used

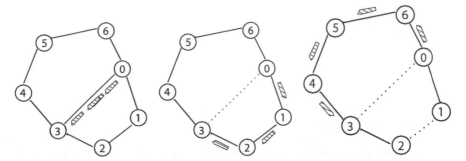

Figure 6.17: DV rout- Figure 6.18: Chooses Figure 6.19: Another ing uses shortest hop new routes when n0 to route found when the path n3 link fails link from n1 to n2 fails again

to design user-defined routing tables. Again, considering the network in Figure 6.11, let us define the path from n0 to n3 as n0→n1→n2→n3, and path from n3 to n0 as n3→n4→n5→n6→n0. The code is given in Listing 6.12. The simulation screenshots are given in Figures 6.20 and 6.21. As can be seen from these figures, the data packets traverse the path shown in Figure 6.20 and the ACKs traverse a separate path, as given in Figure 6.21.

Listing 6.12: Manual routing simulation

```
1  set ns [new Simulator]
2  $ns color 1 blue
3
4  #routing protocol
5  $ns rtproto Manual
6
7  #creating nodes
8  for  {set i 0} {$i < 7} {incr i} {
9    set n($i) [$ns node]
10 }
11
12 set nf [open out.nam w]
13 $ns namtrace-all $nf
14
15 #creating links
16 for {set i 0} {$i < 7} {incr i} {
17   $ns duplex-link $n($i) $n([expr ($i+1)%7]) 1Mb 10ms
       DropTail
18 }
19 $ns duplex-link $n(0) $n(3) 1Mb 10ms DropTail
20
21 #Manual Routes
22 #Path from n0-n3: n0->n1->n2->n3
23 [$n(0) get-module "Manual"] add-route-to-adj-node -default $n
      (1)
```

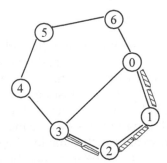
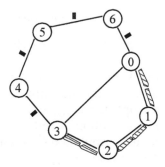

Figure 6.20: Manual routing uses   Figure 6.21: ACK follows a differ-
the path as defined in the program   ent path from the data packets

```
24 [$n(1) get-module "Manual"] add-route-to-adj-node -default $n
      (2)
25 [$n(2) get-module "Manual"] add-route-to-adj-node -default $n
      (3)
26
27 #Path from n3-n0: n4->n5->n6->n0
28 [$n(3) get-module "Manual"] add-route-to-adj-node -default $n
      (4)
29 [$n(4) get-module "Manual"] add-route-to-adj-node -default $n
      (5)
30 [$n(5) get-module "Manual"] add-route-to-adj-node -default $n
      (6)
31 [$n(6) get-module "Manual"] add-route-to-adj-node -default $n
      (0)
32
33 set tcp [new Agent/TCP]
34 $tcp set fid_ 1 ;#blue packets
35 $ns attach-agent $n(0) $tcp
36 set sink [new Agent/TCPSink]
37 $ns attach-agent $n(3) $sink
38 $ns connect $tcp $sink
39 set ftp [new Application/FTP]
40 $ftp attach-agent $tcp
41
42 $ns at 0.5 "$ftp start"
43 $ns at 4.5 "$ftp stop"
44
45 $ns at 5.0 "finish"
46 proc  finish {} {
47    global ns nf
48    $ns flush-trace
49    close $nf
50    puts "running nam..."
51    exec nam out.nam &
52    exit 0
53 }
54
55 $ns run
```

(e) *Other Mechanisms for Specialized Routing*

    i) *Asymmetric Routing:* This routing is required when the path from
one node to another node becomes different while sending from either side that is, if the path computed from node n1 to node n2 is
different from the path from n2 to n1. The code in Listing 6.13 shows
a simple topology, with respective link costs that can result in asymmetric routing. A routing protocol using link costs as the metric for
route computation (link state) may exhibit such behavior if the link
costs are so configured. In Figure 6.22 it can be seen that the traffic
from n0 to n2 and the traffic from n2 to n0 follow different paths
due to asymmetric routing.

Listing 6.13: Asymmetric routing simulation

```
1  set ns [new Simulator]
2
3  $ns color 1 red
4  $ns color 2 blue
5
6  #creating nodes
7  for {set i 0} {$i < 4} {incr i} {
8    set n($i) [$ns node]
9  }
10
11 #trace files
12 set f [open out.tr w]
13 $ns trace-all $f
14 set nf [open out.nam w]
15 $ns namtrace-all $nf
16
17 #creating links
18 for {set i 0} {$i < 4} {incr i} {
19   $ns duplex-link $n($i) $n([expr ($i+1)%4]) 1Mb 10ms
       DropTail
20 }
21
22 $ns duplex-link-op $n(0) $n(1) label ":2"
23 $ns duplex-link-op $n(1) $n(2) label ":3"
24 $ns duplex-link-op $n(2) $n(3) label ":2"
25
26 #routing protocol
27 $ns rtproto Session
28
29 #Assign costs to links
30 $ns cost $n(0) $n(1) 2
31 $ns cost $n(1) $n(2) 3
32 $ns cost $n(2) $n(3) 2
33
34 #Create a UDP agent and a sink and attach them
35 #node n(0) and n(2) respectively
36 set udp0 [new Agent/UDP]
37 $ns attach-agent $n(0) $udp0
38 set null0 [new Agent/Null]
```

```
39  $ns attach-agent $n(2) $null0
40  $udp0 set fid_ 1
41  set cbr0 [new Application/Traffic/CBR]
42  $cbr0 set packetSize_ 500
43  $cbr0 set interval_ 0.009
44  $cbr0 attach-agent $udp0
45  $ns connect $udp0 $null0
46
47  #Create another UDP agent and a sink, then attach them to
        node n(2) and to n(0) respectively
48  set udp1 [new Agent/UDP]
49  $ns attach-agent $n(2) $udp1
50  set null1 [new Agent/Null]
51  $ns attach-agent $n(0) $null1
52  $udp1 set fid_ 2
53  set cbr1 [new Application/Traffic/CBR]
54  $cbr1 set packetSize_ 500
55  $cbr1 set interval_ 0.007
56  $cbr1 attach-agent $udp1
57  $ns connect $udp1 $null1
58
59  $ns at 0.5 "$cbr0 start"
60  $ns at 4.5 "$cbr0 stop"
61  $ns at 0.7 "$cbr1 start"
62  $ns at 4.7 "$cbr1 stop"
63
64  $ns at 0.0 "$udp0 set fid_ 1"
65  $ns at 0.0 "$udp1 set fid_ 2"
66
67  $ns at 5.0 "finish"
68  proc finish {} {
69    global ns nf
70    $ns flush-trace
71    close $nf
72    puts "running nam..."
73    exec nam out.nam &
74    exit 0
75  }
76
77  $ns run
```

ii) **Multipath Routing:** A node may be configured to use multiple different paths to a particular destination. If multiple paths to a destination from a source are available, then that source node can use all of the different paths to transmit packets to the destination simultaneously.

node set multiPath_ 1

All the nodes present in the simulation now are capable of using multiple paths wherever possible or may use each of them alternately.

$n1 set multiPath_ 1

This command enables only node n1 to use multiple paths if applicable. In NS only distance vector routing can generate multipath

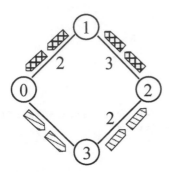

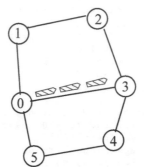

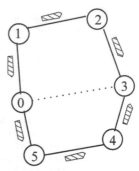

Figure 6.22: Asymmetric path routing

Figure 6.23: Multipath routing

Figure 6.24: Using multiple paths (n0-n3 breaks)

routes. In Figure 6.23 the traffic uses the shortest hop path, but when the link between n0 and n3 fails, the routing agent has two paths with equal hop counts and the nodes are multipath enabled, so it uses both the paths (n0→n1→n2→n3 and n0→n5→n4→n3), as shown in Figure 6.24.

Listing 6.14: Multipath routing simulation

```
1  set ns [new Simulator]
2  Node set multiPath_ 1
3
4  #creating nodes
5  for {set i 0} {$i < 6} {incr i} {
6      set n($i) [$ns node]
7  }
8
9  set nf [open out.nam w]
10 $ns namtrace-all $nf
11
12 #creating links
13 for {set i 0} {$i < 6} {incr i} {
14     $ns duplex-link $n($i) $n([expr ($i+1)%6]) 1Mb 10ms
           DropTail
15 }
16
17 $ns duplex-link $n(0) $n(3) 1Mb 10ms DropTail
18
19 #routing protocol
20 $ns rtproto DV
21
22 #Create a UDP agent and attach it to node n(0)
23 set udp0 [new Agent/UDP]
24 $ns attach-agent $n(0) $udp0
25
26 # Create a CBR traffic source and attach it to udp0
27 set cbr0 [new Application/Traffic/CBR]
```

```
28  $cbr0 set packetSize_ 500
29  $cbr0 set interval_ 0.01
30  $cbr0 attach-agent $udp0
31  set null0 [new Agent/Null]
32  $ns attach-agent $n(3) $null0
33  $ns connect $udp0 $null0
34
35  $ns at 0.5 "$cbr0 start"
36  $ns at 4.5 "$cbr0 stop"
37
38  #link failure
39  $ns rtmodel-at 1.5 down $n(0) $n(3)
40  $ns rtmodel-at 2.5 up $n(0) $n(3)
41  $ns rtmodel-at 2.0 down $n(1) $n(2)
42  $ns rtmodel-at 3.5 up $n(1) $n(2)
43
44  $ns at 5.0 "finish"
45  proc finish {} {
46    global ns nf
47    $ns flush-trace
48    close $nf
49    puts "running nam..."
50    exec nam out.nam &
51    exit 0
52  }
53  $ns run
```

Test/Viva Questions

a) What is routing? Which layer of the ISO/OSI protocol performs this function?

b) What are different routing strategies available in NS?

c) Which routing protocol(s) use Dijkstra's all-pairs SPF algorithm?

d) Which routing protocol(s) use the distributed Bellman–Ford algorithm?

e) What are the advantages/disadvantages of static routing?

f) Compare and contrast static, dynamic, and manual routing strategies.

g) Why is the *split horizon with poisoned reverse* used in the distance vector routing scheme?

h) Compare session and distance vector routing strategies.

i) What is asymmetric routing?

j) What is multipath routing? Which routing protocol does it use?

Programming Assignments

1) Construct a network as given in Figure 6.25. Apply distance vector routing and session routing to find a route between node 0 and node 4. Then compare/justify these two protocols with respect to the path selected and the average delay incurred by the network.

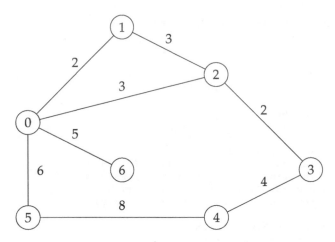

Figure 6.25: Seven-node network

6.6   Transport Layer — Transport Agents

There are many transport agents available in NS2. We restrict our discussion to the two fundamental ones, UDP and TCP.

6.6.1   User datagram protocol (UDP)

The UDP agent provides a connectionless unreliable service in the transport layer. From the application layer, the UDP agent receives data in chunks and creates packets from it. Unlike the standard UDP protocol, UDP packets in NS use a monotonically increasing sequence number and a time stamp. This sequence number does not impose any simulation overhead, but is an important requirement for trace file analysis and performance measurement, discussed in Section 6.7.

Commands

UDP agent instance creation

```
set udp0 [new Agent/UDP]
```

#Attach the UDP agent to a any node

```
$ns attach-agent <node >  <agent>
```

#To join a source agent with a sink agent (not a connection establishment, but fixing a source and a corresponding destination)

```
$ns_ connect <src-agent> <dst-agent>
```

UDP configuration parameters

```
$udp set packetSize_ <pktsize> ;# default 1000 bytes
$udp set dst_addr_ <address>
$udp set dst_port_ <portnum>
$udp set class_ <class-type>
```

```
$udp set ttl_ <time-to-live>
```
... etc.

The default values for the above parameters may be found in ~ns/tcl/lib/ns-default.tcl.

Almost all programs illustrated so far use UDP at the transport layer, so no separate examples are provided here. But the reader should use UDP with the varying value of parameters given above and compare the results.

6.6.2 Transmission control protocol (TCP)

This is undoubtedly the most popular transport protocol that provides connection-oriented, reliable transmission service. In NS2 there are two major categories of TCP agents available, as discussed below.

i. **One-Way Agents**: These agents are agents that can either send or receive but cannot perform both operations. The term sending agent refers to a source agent and receiving agent refers to the sink agent. These agents are further subdivided into a set of TCP senders according to their congestion handling and error control characteristics.

ii. **Two-Way Agent**: This agent is symmetric, that is, unlike the one-way agent, the same agent can be used as both a sender and a receiver simultaneously.

Commands

\# Create TCP agent and attach it to the node
```
set tcp0 [new Agent/TCP]
$ns attach-agent $n0 $tcp0
```
\#Create a receiver agent and attach it to the node
```
set sink0 [new Agent/TCPSink]
$ns attach-agent $n1 $sink0
```
\#Connect the sender and receiver agents
```
$ns connect $tcp0 $sink0
```

Different implementations of TCP agents attempt only to achieve the objectives of real-world TCP characteristics, but are not exact implementations of real-world TCP. For example, the following differences exist.

i. Unlike the real TCP protocol that uses window advertisements from the receiver's side, NS TCP agents do not contain a dynamic window advertisement, as no data is ever transferred.

ii. ACK number computations are done entirely in packet units, unlike real-world TCP, where segment numbers are computed as the first byte number of the segment.

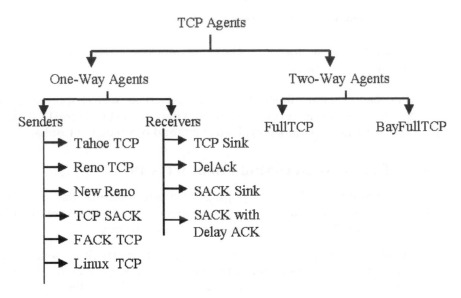

Figure 6.26: TCP variants in NS

iii. There are no SYN/FIN segments used for connection establishment or teardown, and no checksums or urgent data, etc.

The different implementations of TCP in NS are provided in Figure 6.26. Some of them are discussed below.

One-Way TCP sending Agents

- **Tahoe TCP Sender** (Agent/TCP)
  This is the base TCP implementation in NS. It implements 4.3BSD Tahoe Unix system release from UC Berkeley. Some of the features are as follows.

  i) During the slow-start phase, the congestion window ($cwnd\_$) is increased by one packet per new ACK received. That is, when the congestion window is smaller, then the slow-start threshold ($cwnd\_ < ssthresh\_$) $cwnd\_$ is increased by one packet. Similarly, during the congestion avoidance phase ($cwnd\_ \geq ssthresh\_$) the $cwnd\_$ is increased by $\frac{1}{cwnd\_}$ for each new ACK received.

  ii) The congestion in a network is understood when TCP observes either (i) a retransmission timer expires or (ii) three duplicate ACKs are received. In either case, Tahoe TCP reacts by setting the slow-start threshold ($ssthresh\_$) to $max\left(\frac{window\_}{2}, 2\right)$ and the current window size ($window\_$) is considered as the $min(cwnd\_, window\_)$. It then initializes $cwnd\_$ back to the value of $windowInit\_$, which causes the TCP to enter into the slow-start phase. The default values for each of these parameters are provided below.

iii) The sample for RTT is calculated by finding the difference between the current time and a time stamp of the ACK packet. The initial value for *srtt_* is set to the value of the first sample. Similarly, the initial value of *rttvar_* is computed as half of the first sample (*srtt_*). For subsequent samples, the values are updated as follows:

$$srtt_ = \frac{7}{8} \times srtt_ + \frac{1}{8} \times sample$$
$$rttvar_ = \frac{3}{4} \times rttvar_ + \frac{1}{4} \times |sample - srtt_|$$

Configuration Parameters

`Agent/TCP set window_ 100` ;# Changes the class variable

`$tcp set window_ 2.0` ;# Changes window_ for the $tcp object only

The default values for different parameters of the TCP agent are given as follows.

Parameters	Values	Use
window_	20	max. current window size
windowInit_	1	initial value of congestion window
windowOption_	1	congestion avoidance algorithm
windowConstant_	4	used only when windowOption !=1
windowThresh_	0.002	used in computing averaged window
overhead_	0	if !=0, then adds random time between sends
ecn_	0	TCP should not react to explicit congestion notation bit
packetSize_	1000	packet size used by sender in bytes
slow_start_restart_	true	
tcpTick_	0.1	timer granularity in sec
dupacks_	0	duplicate ACK counter
ack_	0	highest ACK received 294
cwnd_	0	congestion window (packets)
awnd_	0	averaged cwnd (experimental)
ssthresh_	0	slow-start threshold (packets)

`rtt_`	0	rtt sample
`srtt_`	0	smoothed (averaged) rtt
`rttvar_`	0	mean deviation of rtt samples
`backoff_`	0	current RTO backoff factor
`maxseq_`	0	max (packet) seq number sent

- **Reno TCP sender** (Agent/TCP/Reno)
  The Reno TCP agent is similar to the Tahoe TCP agent but includes *fast recovery* and *fast retransmit* mechanisms. In *fast recovery*, the current congestion window is reduced to half its size and *ssthresh_* is reset to match this value, without returning to slow start. In *fast retransmit*, the sender triggers retransmission when three duplicate ACK arrive, without waiting for a timeout.

- **New Reno TCP sender** (Agent/TCP/Newreno)
  This is essentially Reno with a modification. This agent is based on the Reno TCP agent with one difference. The sending agent exits from fast recovery when it receives an ACK for the latest segment sent. New "partial ACKs" representing new ACKs but not representing an ACK for all outstanding data do not cause any shrinkage in the window size.

- **TCP SACK Sender** (Agent/TCP/Sack1)
  TCP with selective repeat (RFC2018). The receiver sends selective ACKs, and the agent implements selective repeat based on the ACK provided by the receiver.

- **Linux TCP Sender** (Agent/TCP/Linux)
  This is a TCP sender with selective repeat (SACK) that runs TCP congestion control modules available in a Linux kernel. Users may import different congestion control modules available in the Linux kernel source code. The Linux congestion control modules are now compiled into the NS binary.

  Simulation results close to Linux performance may be achieved by changing the default values of the following parameters according to the Linux parameters as given below

  ```
  Agent/TCP/Linux set maxrto_ 120

  Agent/TCP/Linux set ts_resetRTO_ true

  Agent/TCP/Linux set delay_growth_ false
  ```

One-Way TCP Receiving Agents

- **TCP sink with one ACK per packet** (Agent/TCPSink)
  This is the base TCP receiver (sink) object that is responsible for returning ACKs to the peer TCP sending object. It generates one ACK for each packet received. ACK size may be configured as follows.
  Configuration Parameters

  Agent/TCPSink set packetSize_ 40

- **TCP sink with configurable delay per ACK** (Agent/TCPSink/DelAck)

  Unlike a base TCP receiver object, this is a delayed-ACK receiver object that does not send an ACK for each packet received. It defines a minimum time gap between two ACKs represented by *interval_* variable that provides the number of seconds to wait before sending the next ACK. However, only ACKs for in-order packets are delayed, whereas an immediate ACK is generated for out-of-order packets.
  Configuration Parameters

  emphAgent/TCPSink/DelAck set interval_ 100ms

- **Selective ACK sink** (Agent/TCPSink/Sack1)
  The SACK TCP sink implements the process of selective ACK generation as provided in RFC 2018. A variable *maxSackBlocks_* is used to provide the maximum number of blocks in an ACK that are available for holding information regarding SACK. The value of this variable may be configured as follows and the default value of this variable is 3.
  Configuration Parameters

  Agent/TCPSink set maxSackBlocks_ 5

- **Sack1 with DelAck** (Agent/TCPSink/Sack1/DelAck)
  In this object both Delayed and SACK mechanisms together are implemented.

Now let us try to develop a few simulation programs using TCP agents with some of the techniques discussed above. As we discussed in theory, two flow control algorithms can be used: stop-and-wait and sliding window. In the first flow control algorithm, each time a packet is sent by the sender, it waits for the ACK before sending the next packet. This is the most reliable but inefficient protocol in the sense that sender wastes time waiting for the ACK, resulting in underutilized link bandwidth. The second one uses the concept of a virtual window over the datastream that contains some bytes/packets inside it (normally more than one) and can send a window of packets without waiting for the ACK, hence increasing the use of link bandwidth. In practice, TCP uses the sliding window algorithm. However, the stop-and-wait mechanism can be simulated as a special case of a sliding window by making the window sizeable to contain one packet.

Consider Listing 6.15, where only two nodes are created in lines 5 and 6. The nodes are then labeled as sender and receiver in lines 8 and 9. A link is established between them with a queue size 10 in lines 14 and 15. In line 17 a TCP sending agent is created with maximum window size set to one in lines 18 and 19, thereby forcing TCP to behave like a stop-and-wait mechanism. The rest of the program follows the routine steps, such as attaching agents to nodes, and an ftp application is attached at the top of TCP.

When the program is executed, the sender will send one packet, be quiet until the ACK for the sent packet is received, and continue to send the next packet.

Listing 6.15: Stop-and-wait protocol

```
1  #Stop-and-wait as a special case of Sliding Window
2  #with windowSize and maxWindowSize=1
3  set ns [new Simulator]
4
5  set n0 [$ns node]
6  set n1 [$ns node]
7
8  $ns at 0.0 "$n0 label Sender"
9  $ns at 0.0 "$n1 label Receiver"
10
11 set nf [open stopNWait.nam w]
12 $ns namtrace-all $nf
13
14 $ns duplex-link $n0 $n1 0.2Mb 200ms DropTail
15 $ns queue-limit $n0 $n1 10
16
17 set tcp [new Agent/TCP]
18 $tcp set window_ 1
19 $tcp set maxcwnd_ 1
20 $ns attach-agent $n0 $tcp
21 set sink [new Agent/TCPSink]
22 $ns attach-agent $n1 $sink
23 $ns connect $tcp $sink
24 set ftp [new Application/FTP]
25 $ftp attach-agent $tcp
26
27 $ns at 0.1 "$ftp start"
28 $ns at 3.0 "$ns detach-agent $n0 $tcp ; $ns detach-agent $n1
       $sink"
29
30 $ns at 3.5 "finish"
31 proc finish {} {
32    global ns nf
33    $ns flush-trace
34    close $nf
35    puts "running nam..."
36    exec nam stopNWait.nam &
37    exit 0
38 }
39
40 $ns run
```

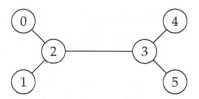

Figure 6.27: Six-node network

This program demonstrates the normal stop-and-wait protocol in ideal conditions, that is, when there is no loss of packets. Also we can associate the packet loss situation, where the sender has to wait for a time out to resend the lost packet. To accommodate packet loss, add the following two commands in Listing 6.15 between lines 27 and 28.

```
$ns at 1.3 "$ns queue-limit $n0 $n1 0"
$ns at 1.5 "$ns queue-limit $n0 $n1 10"
```

We reduce the queue size of the link to 0 in the first command, such that the next packet arrives from the sender at time 1.3 sec, finds no place in the queue and is dropped. Again at time 1.5 sec we increase the queue size to 10 in the second command so that the normal situation is resumed.

When the program is executed after adding the above two commands, packet loss occurs at time 1.3. The sender now waits until the time out of the retransmission timer, after which normal transmission resumes again.

Similarly, the sliding window concept may be simulated by replacing lines 18 and 19 of Listing 6.15 as follows.

```
11. $tcp set windowInit_ 4
12. $tcp set maxcwnd_ 4
```

The above commands increase the window size to four packets, that is, TCP sender can now send four packets without waiting for the ACK. When executed, the TCP sending agent will send four packets at a time and also receive four ACKs from the sink agent, and the process is repeated until the end of the simulation.

Now let's try to write a more interesting program that demonstrates practical TCP behavior in different situations like overload conditions or congestion situations. The behavior can be observed by measuring the congestion-window size variation and throughput at different time intervals of the simulation. To simulate the above ideas, consider the network given in Figure 6.27. In this network node 0 sends ftp traffic over TCP to node 4. In Listing 6.16, lines 3 and 4 create an NS trace and a NAM trace, respectively. Line 5 creates a file called cwnd.dat to record the congestion window size at different time instants. Lines 10 to 17 define the finish procedure that is executed when the simulation ends. Lines 19 to 21 create six nodes, and lines 23 to 27 create links between them. Line 25 creates the bottleneck at node 2. This is because node 0 and node 1 can push data at a rate of 1 Mbps, whereas node

2 can forward 0.1 Mbps data. Lines 28 to 32 provide link orientations. The network is arranged like Figure 6.27 in NAM visualization. Line 33 is to visualize the queue status at node 2, and line 35 defines the queue size at node 2 as 16 packets.

Lines 37 to 42 are used to create the TCP agent and attach to source node 0 and the TCP sink agent to node 4. Lines 43 to 46 attach the ftp traffic to the TCP sending agent and packet transmission start and end timings. Lines 48 to 55 define a procedure recWin to record the dynamic congestion window size into the file cwnd.dat in the time interval of 0.1 sec. Line 57 invokes the recWin procedure for the first time at 0.1 sec, and the procedure is called recursively in line 54 at each 0.1 sec time interval. In other words, the window size is recorded at times $0.1, 0.2, 0.3, \ldots$. This granularity may be changed by changing the value 0.1 in lines 50 and 57 to some other value. Finally, line 58 defines the total simulation time (124 sec), and line 60 runs the simulation.

Listing 6.16: TCP behavior

```
1  #Transport Layer Simulation
2  set ns [new Simulator]
3  set tr1 [open tcpWin.tr w]
4  set nam1 [open tcpWin.nam w]
5  set cwnd [open cwnd.dat w]
6
7  $ns trace-all $tr1
8  $ns namtrace-all $nam1
9
10 proc  finish {} {
11    global ns tr1 nam1
12    $ns flush-trace
13    close $tr1
14    close $nam1
15    exec nam tcpWin.nam &
16    exit 0;
17 }
18
19 for {set i 0} {$i < 6 } {incr i} {
20    set n$i [$ns node]
21 }
22
23 $ns duplex-link $n0 $n2 1Mb 10ms DropTail
24 $ns duplex-link $n1 $n2 1Mb 10ms DropTail
25 $ns duplex-link $n2 $n3 0.1Mb 100ms DropTail
26 $ns duplex-link $n3 $n4 0.5Mb 50ms DropTail
27 $ns duplex-link $n3 $n5 0.5Mb 50ms DropTail
28 $ns duplex-link-op $n0 $n2 orient right-down
29 $ns duplex-link-op $n1 $n2 orient right-up
30 $ns duplex-link-op $n2 $n3 orient right
31 $ns duplex-link-op $n3 $n4 orient right-up
32 $ns duplex-link-op $n3 $n5 orient right-down
33 $ns duplex-link-op $n2 $n3 queuePos 0.5
34
```

```
35  $ns queue-limit $n2 $n3 10
36
37  set tcp [new Agent/TCP]
38  $ns attach-agent $n0 $tcp
39  set sink [new Agent/TCPSink]
40  $ns attach-agent $n4 $sink
41  $ns connect $tcp $sink
42  $tcp set packetSize_ 2048
43  set ftp [new Application/FTP]
44  $ftp attach-agent $tcp
45  $ns at 1.0 "$ftp start"
46  $ns at 120.0 "$ftp stop"
47
48  proc recWin {tcpSender f} {
49     global ns
50     set time 0.1
51     set cTime [$ns now]
52     set wnd [$tcpSender set cwnd_]
53     puts $f "$cTime $wnd"
54     $ns at [expr $cTime+$time] "recWin $tcpSender $f"
55  }
56
57  $ns at 0.1 "recWin $tcp $cwnd"
58  $ns at 124 "finish"
59
60  $ns run
```

Table 6.1: TCP window size recorded in file cwnd.dat

0.10000000000000001	1
0.20000000000000001	1
. . .	. . .
1.4000000000000001	2
1.5000000000000002	2
. . .	. . .
5.0999999999999979	20.05
5.1999999999999975	20.05
. . .	. . .
8.1999999999999869	20.8336
8.2999999999999865	1
8.3999999999999861	1
. . .	. . .

When Listing 6.16 is executed, it generates three files, out of which the NAM trace file (tcpWin.nam) is executed automatically to produce the animation when the simulation ends due to line 15. The NS trace file (tcpWin.tr) records all the events of simulation, and the third file (cwnd.dat) records the dynamic congestion window size in packets at each 0.1 sec. Table 6.1 gives few lines of cwnd.dat, where column 1 represents the time of recording and column 2 is the window size at that time. To observe TCP behavior more

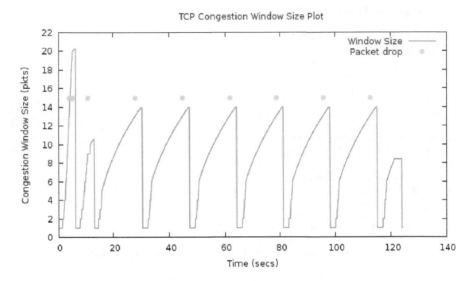

Figure 6.28: Congestion window size variation

closely, we can plot this information using Gnuplot. To do this, issue the following commands.

Gnuplot

plot "cwnd.dat" with lines

This produces a graph showing the variation of window size. To make this plot more meaningful, it may be customized by writing a Gnuplot script as given in Listing 6.17 in a file winplot.p. To execute the script, issue the command

Gnuplot

load 'winplot.p'

These commands will produce the plot shown in Figure 6.28.

Listing 6.17: Gnuplot script to plot congestion window

```
1  #file name: winplot.p
2
3  set autoscale
4  unset log
5  unset label
6  set xtic auto
7  set ytic auto
8  set title "TCP Congestion Window Size Plot"
9  set xlabel "Time (secs)"
10 set ylabel "Congestion Window Size (packets)"
11
12  plot "cwnd.dat" using 1:2 title 'QueueSize:10' with lines
```

Observe Figure 6.28, in which, after a packet drop is experienced, the congestion window size drops to 1 packet according to the slow-start mechanism and then resumes the window size by rapidly increasing the window size up to half of the congestion threshold, after which the congestion window size increases slowly until another packet drop occurs. This is called additive increase.

Again we can also calculate the variation in throughput at different time instants of TCP by using an AWK script, as given in Listing 6.18. This AWK script extracts information from the trace file regarding the packets received at a destination in a particular time interval and calculates the throughput at that time interval. To execute the AWK script, issue the command

```
awk -f thput.sh tcpWin.tr > thput
```

Listing 6.18: AWK script to find throughput

```
1  #!\bin\awk -f
2  # file name: thput.sh
3
4  BEGIN {
5    size = 0;
6    interval = 1;
7    dest = 4;
8    Time = 0;
9    throughput = 0;
10 }
11 {
12   event = $1;
13   time = $2;
14   from = $3;
15   to = $4;
16   pktType = $5;
17   pktSize = $6;
18   if (time-Time <= interval) {
19     if (event == "r") {
20       if ( to == dest) {
21         if ( pktType == "tcp" ) {
22           size += pktSize;
23         }
24       }
25     }
26   }
27   else {
28     throughput = size*8 / interval;
29               throughput /= 1000;
30     printf time "\t" throughput "\n"
31     Time = Time+interval;
32     size = 0;
33   }
34 }
```

This command will calculate throughput and save in a file called thput. A few lines of this file a shown in Table 6.2, in which the first column represents time in seconds and the second column represents throughput in Kbps. The throughput is measured in each 1 sec time interval. The plot will be more explanatory if packet drops are shown along with throughput. To find the packet drops, write another AWK script, as shown in Listing 6.19.

Table 6.2: TCP throughput in file thput

1.01032	0
2.026352	33.728
3.065616	83.52
4.000816	100.224
. . .	. . .
8.117008	83.52
9.005888	50.112
10.008128	100.224
. . .	. . .

Listing 6.19: AWK script to find packet drops

```
1  #File name: pktdrop.sh
2  #!\bin\ awk -f
3  {
4    event = $1;
5    time = $2;
6    if (event == "d") {
7      printf time "\t" "102" "\n"
8    }
9  }
```

Execute and store the packet drop information in a file called pktdrops. To plot the throughput, write another Gnuplot script, as given in Listing 6.20.

Listing 6.20: Gnuplot script to plot variation of throughput

```
1  #file name: thput.p
2  set autoscale
3  unset log
4  unset label
5  set xtic auto
6  set ytic auto
7  set title "TCP Throughput at Destination in different Time
     Intervals"
8  set xlabel "Simulation Time (secs)"
9  set ylabel "Throughput (kbps)"
10
11  plot "pktdrops" u 1:2 title 'Packet Drop' w p pt 7, "thput" u
     1:2 title 'Throughput at 2' w l lt 2
```

This script produces the plot shown in Figure 6.29. Packet drops are shown as big dots and throughput is shown as lines. As can be seen, the throughput drops suddenly after a packet drop and regains throughput after some time. The average throughput over total time may calculated by changing the 'else' part of Listing 6.18 as follows.

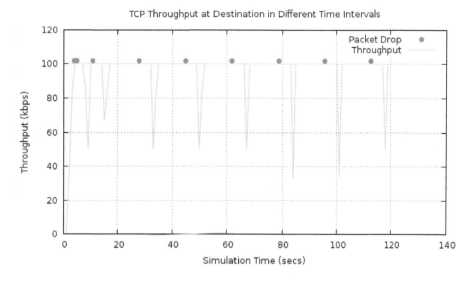

Figure 6.29: Throughput variation

```
else{
    if(time > 0){
        throughput - size*8 / time;
        throughput /= 1000 ;
        printf time "\t" throughput "\n";
    }
    Time += interval;
}
```

The figures for throughput along with packet drop timings are shown in Figures 6.29 and 6.30.

Two-Way Experimental Sender

- **Full TCP** (Agent/TCP/FullTcp)
  Full TCP currently supports a Reno form of TCP. It is different and hence incompatible with the other one-way agents discussed above. In this case we do not need to use different agents for sender and receiver. The characteristics of this agent that distinguish it from other agents are as follows.

    - Connections are *established* before data transfer and *teared down* after finishing data transfer. (SYN/FIN packets are exchanged)

    - Sending and receiving (duplex communication) by the same agent is possible.

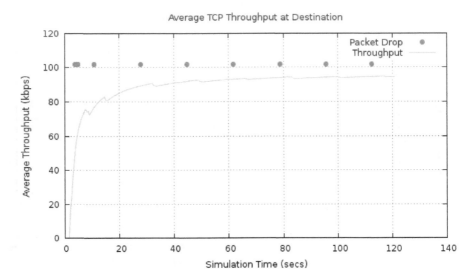

Figure 6.30: Average TCP throughput

- Sequence numbers are provided for each *byte* received from the application, unlike one-way agents, where a sequence number is assigned to the packets.

This TCP agent sends data on the third segment of an initial three-way handshake process, which is somewhat earlier than typical real-world implementations. This is because the practical TCP implementations typically follow the following pattern (three-way handshaking).

- Client TCP (active opener) sends a *SYN* segment.
- Server TCP (passive opener) responds back with a *SYN+ACK* segment.
- The active opener now responds with an *ACK*, which completes the connection establishment phase.
- Some time later the first data segment is sent.

Steps to Use Full TCP

```
set src [new Agent/TCP/FullTcp] ;# create agent
set sink [new Agent/TCP/FullTcp] ;# create agent
$ns_ attach-agent $node_(s1) $src ;# bind src to node
$ns_ attach-agent $node_(k1) $sink ;# bind sink to node
$src set fid_ 0 ;# set flow ID field
$sink set fid_ 0 ;# set flow ID field
```

$ns_ connect $src $sink ;# active connection src to sink
set up TCP-level connections

$sink listen ;# will figure out who its peer is

$src set window_ 100;

As can be observed from the above steps, the creation of the FullTcp agent is similar to the other agents, but the sink is placed in a listening state by the listen method because a handle to the receiving side is required in order to make this call.

Configuration Parameters
segs received before generating ACK

> **Agent/TCP/FullTcp set segsperack_ 1**
segment size (MSS size for bulk transfers)

> **Agent/TCP/FullTcp set segsize_ 536**
dupACKs thresh to trigger fast retransmit

> **Agent/TCP/FullTcp set tcprexmtthresh_ 3**
initial send sequence number

> **Agent/TCP/FullTcp set iss_ 0**
disable sender-side Nagle algorithm

> **Agent/TCP/FullTcp set nodelay_ false**
send data on initial SYN?

> **Agent/TCP/FullTcp set data_on_syn_ false**
avoid fast rxt due to dup segs+acks

> **Agent/TCP/FullTcp set dupseg_fix_ true**
reset dupACK ctr on non-zero len data segs containing dup ACKs

> **Agent/TCP/FullTcp set dupack_reset_ false**
Packet interval

> **Agent/TCP/FullTcp set interval_ 0.1**

A program using full TCP is given in Listing 6.21. This program also assumes the same topology as that given in Figure 6.27, but the transmission occurs in both directions, that is, full duplex transmission between node 0 and node 4. Lines 36 and 38 create two TCP agents and are attached to node 0 and node 4, respectively, unlike one-way sending TCP, where one TCP agent and one sink are required. The data transmission can now be done from both sides. Execute the simulation and observe the duplex transmission between two nodes in NAM visualization.

Listing 6.21: Full TCP simulation

```
1 #Transport Layer Simulation with Full TCP
```

```
2  set ns [new Simulator]
3  set tr1 [open winTcp.tr w]
4  set nam1 [open winTcp.nam w]
5
6  $ns trace-all $tr1
7  $ns namtrace-all $nam1
8
9  proc  finish {} {
10   global ns tr1 nam1
11   $ns flush-trace
12   close $tr1
13   close $nam1
14   exec nam winTcp.nam &
15   exit 0;
16 }
17
18 for {set i 0} {$i < 6 } {incr i} {
19   set n$i [$ns node]
20   }
21
22 $ns duplex-link $n0 $n2 1Mb 10ms DropTail
23 $ns duplex-link $n1 $n2 1Mb 10ms DropTail
24 $ns duplex-link $n2 $n3 0.1Mb 100ms DropTail
25 $ns duplex-link $n3 $n4 0.5Mb 50ms DropTail
26 $ns duplex-link $n3 $n5 0.5Mb 50ms DropTail
27
28 $ns duplex-link-op $n0 $n2 orient right-down
29 $ns duplex-link-op $n1 $n2 orient right-up
30 $ns duplex-link-op $n2 $n3 orient right
31 $ns duplex-link-op $n3 $n4 orient right-up
32 $ns duplex-link-op $n3 $n5 orient right-down
33 $ns duplex-link-op $n2 $n3 queuePos 0.5
34 $ns queue-limit $n2 $n3 10
35
36 set src [new Agent/TCP/FullTcp]
37 $ns attach-agent $n0 $src
38 set sink [new Agent/TCP/FullTcp]
39 $ns attach-agent $n4 $sink
40 $src set fid_ 0
41 $sink set fid_ 1
42 $ns connect $src $sink
43 $sink listen
44 $src set segsize_ 536
45 set ftp1 [new Application/FTP]
46 $ftp1 attach-agent $src
47 set ftp2 [new Application/FTP]
48 $ftp2 attach-agent $sink
49
50 $ns at 1.0 "$ftp1 start"
51 $ns at 2.0 "$ftp2 start"
52 $ns at 120.0 "$ftp2 stop"
53 $ns at 20.0 "$ftp2 stop"
54 $ns at 124 "finish"
55
56 $ns run
```

- **BayFullTcp**
  This is a separate implementation of two-way TCP available in NS. The major differences between BayFullTcp and FullTcp are as follows.

  - *BayFullTcp* supports a client-server application model, while *FullTcp* makes no assumption about its application layer.
  - The TCP-application interface is different for both of them.
  - *FullTcp* supports partial ACK, but *BayFullTcp* does not.
  - *FullTcp* supports different flavors of tcp (tahoe, reno etc. etc), which is not the case for *BayFullTcp*.
  - Both implementations have different set of APIs.

Test/Viva Questions

a) Compare the characteristics of UDP and TCP.

b) Explain the UDP configuration parameters.

c) What is the difference between TCP's one-way and two-way agents?

d) What is *fast recovery*, which TCP Agent uses?

e) How does the new Reno TCP differ from Reno TCP?

f) What is selective repeat? Which TCP agent uses this technique?

g) Justify that the stop-and-wait ARQ is a special case of sliding window ARQ.

h) Name any five TCP configuration parameters and explain their meanings.

i) What is a congestion window? Why is it required?

j) Explain the slow-start technique and its advantages.

Programming Assignments

1) Execute the ideal stop and wait ARQ as in Listing 6.15 and execute again after adding packet loss. Now calculate the throughput of normal stop-and-wait and compare it with the throughput at packet loss conditions.

2) Create a network as given in Figure 6.27. Now attach two TCP agents to node 0 and node 1 and sink agents to nodes 4 and 5, respectively. Make one TCP to follow stop-and-wait ARQ and the other to follow a sliding window of size 8 packets. Now run the program and compare the throughput of different connections. Also find the total throughput of the network and find the ratio of sharing of this throughput by different connections.

3) In the network given in Figure 6.27, attach one TCP agent to node 1. Now measure the congestion window variation for different queue limits for the link 2-3 at different time intervals, that is, 10, 15, 20. Plot the window variation in one plot with proper labeling. Also measure the throughput at different time intervals for the different queue limits given above separately and plot them in one plot and compare them.

4) In the network given in Figure 6.27, attach one UDP agent to node 0 and another TCP agent to node 1. Now measure the congestion window variation at different time intervals for different packet sizes. Plot the window variation in a plot with proper labeling. Also measure the throughput at different time intervals for TCP and UDP separately and plot them in one plot and compare them.

5) In the network given in Figure 6.27, attach two TCP agents to node 0 and node 1 with different packet sizes. Now measure the congestion window variation for different time intervals and plot the window variation of both the TCPs along with packet drops in one plot with proper labeling. Also measure the throughput at different time intervals for both TCPs and plot them in one plot and compare them.

6.7 Packet Trace

As discussed earlier in this chapter, NS can provide two types of output for any simulation. The first one is the NAM trace, which has already been discussed. The second is the packet/event trace, which records each event of the simulation row-wise in a file. Normally, the event trace file's extension is <fileName>.tr. The packet trace information is very important, as it is useful to measure the performance of the network. To effectively extract the required information for various network performance parameter values from a packet trace file is the key challenge here. To do so, the format of the packet trace file must be understood. To generate a packet trace file for any simulation, use the following Tcl code in an NS simulation script, as discussed in Step II of Section 6.2.

```
set ptf [open <pktTrace.tr> w]
$ns trace-all $ptf
```

When the simulation finishes, the trace file named pktTrace.tr is produced that stores all the events generated in the simulation. Each event is recorded in a separate row in the trace file. A few events are stored in a packet trace file as follows.

```
Event 1
+ 1.01 0 1 cbr 500 ----- 0 0.0 1.0 2 2
```

Table 6.3: Event trace format

Column#	Event Trace	Meaning
1	+	Event
2	1.01	Time of Event
3	0	ID of this node
4	1	ID of next hop
5	cbr	Packet type
6	500	Packet size
7	- - - - - - -	Explicit congestion notification (ECN) flag
8	0	IP flow identifier
9	0.0	Source address
10	1.0	Destination address
11	2	Packet sequence number
12	2	Packet unique identifier

Event 2
```
- 1.01 0 1 cbr 500 ----- 0 0.0 1.0 2 2
```
Event 3
```
r 1.014 0 1 cbr 500 ----- 0 0.0 1.0 0 0
...
```
In one trace file you will find many lines as above all these lines have a common format. The information is stored in different columns of a line. So, if we understand one event (one line), then we can extract the required information from multiple events available in a trace file using AWK scripting which was discussed in the last chapter.

The event trace format is provided in Table 6.3, where the first column of the table represents the column number, the second column provides example values from event 1 above, and the last column provides the meaning of that column.

The first column of each line represents the event that occurs. The various events possible are as follows.

```
+: enqueue into the link buffer
-: deque from the link buffer
r: receive by node
d: drop from queue
```
Column 2 provides the time of the event. Columns 3 and 4 tell with which two nodes the event happened. These nodes may be different from the actual source and destination if the traffic flows through intermediate nodes/routers. Column 5 tells the packet type, and column 6 is the packet size in bytes. Column 7 is not used here. Column 8 is the IP flow identifier as used in IPv6. Columns 9 and 10 provide the source and destination

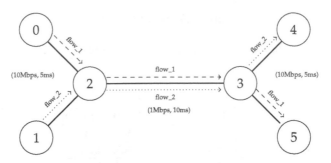

Figure 6.31: A bottle-neck network

addresses. Column 11 is not used in NS2, but is kept to remain backward compatible with NS1. Each packet generated in NS is assigned a new unique identifier and is provided in the last column of the event trace line.

To measure network performance parameters, first the parameters must be defined properly. We define two major parameters, delay and throughput, in the following.

Average Delay
This is the average of total delay incurred due to all packets. The average delay may be calculated as follows.
1. For each packet i received, find the delay as
$delay_i$
$= arrival\ time\ of\ Packet_i\ -\ sending\ time\ of\ Packet_i.$
2. $average\ delay\ =\ \frac{\sum_{i=1}^{n} delay_i}{n}$
This delay is averaged over n received packets.

Network Throughput
This is defined as the total number of bytes received at the destination per second. To calculate this, we can use following two simple steps.
1. Add the bytes of all data packets received.
2. Divide the total size calculated in the above step by the simulation time. (To get the values in Mbps multiply the final value by 8/1,000,000.)

Consider Figure 6.31. We call it a bottleneck network, as the middle link (2-3) has considerably low bandwidth compared to the end-node links (0-2, 1-2, 3-4, and 3-5). Two CBR flows are attached on the top of TCP with identical packet sizes and packet rates. The simulation program is provided in Listing 6.22. Lines 23 to 27 are used to configure different links. The queue associated with each link is of the drop tail type. Lines 32 to 57 are used to attach two CBR traffics; the traffic from $0 \rightarrow 5$ is called $flow\_1$ and similarly the traffic from $1 \rightarrow 4$ is called $flow\_2$. They are red and blue packets. When both flows send packets, the packets are queued at node 2; if the queue becomes full, the next incoming packet is dropped.

In lines 16 to 17, the packet trace is created and the trace is written to a file called traceExp.tr. When the program is executed, it creates both the packet trace (traceExp.tr) and NAM trace (traceExp.nam). The NAM trace is used for visualization and it automatically starts after execution due to line 68 of the program. When the NAM finishes, open the traceExp.tr file in any text editor and find that the event traces are reported as discussed earlier in this section. The complete format is provided in Table 6.3.

Listing 6.22: Bottleneck network

```
1  # Event trace Example
2  #Creating event Scheduler
3  set ns [new Simulator]
4  $ns color 1 red
5  $ns color 2 blue
6
7  #Creating six Nodes
8  set n0 [$ns node]
9  set n1 [$ns node]
10 set n2 [$ns node]
11 set n3 [$ns node]
12 set n4 [$ns node]
13 set n5 [$ns node]
14
15 # Event Trace
16 set tf [open traceExp.tr w]
17 $ns trace-all $tf
18 #Network Animator
19 set nf [open traceExp.nam w]
20 $ns namtrace-all $nf
21
22 #creating link between two nodes
23 $ns duplex-link $n0 $n2 10Mb 5ms DropTail
24 $ns duplex-link $n1 $n2 10Mb 5ms DropTail
25 $ns duplex-link $n2 $n3 1Mb 10ms DropTail
26 $ns duplex-link $n3 $n4 10Mb 5ms DropTail
27 $ns duplex-link $n3 $n5 10Mb 5ms DropTail
28 $ns queue-limit $n2 $n3 20
29 $ns duplex-link-op $n2 $n3 queuePos 0.5
30
31 #UDP Agent between n0 and n5 (flow_1)
32 set udp0 [new Agent/UDP]
33 set null0 [new Agent/Null]
34 $ns attach-agent $n0 $udp0
35 $ns attach-agent $n5 $null0
36 $ns connect $udp0 $null0
37 $udp0 set fid_ 1
38
39 #UDP Agent between n1 and n4 (flow_2)
40 set udp1 [new Agent/UDP]
41 set null1 [new Agent/Null]
42 $ns attach-agent $n1 $udp1
```

```
43  $ns attach-agent $n4 $null1
44  $ns connect $udp1 $null1
45  $udp1 set fid_ 2
46
47  # CBR traffic between n0 and n5 (Red)
48  set cbr [new Application/Traffic/CBR]
49  $cbr   set packetSize_ 500
50  $cbr   set interval_ 0.005
51  $cbr attach-agent $udp0
52
53  # CBR traffic between n1 and n4 (Blue)
54  set cbr1 [new Application/Traffic/CBR]
55  $cbr1 set packetSize_ 500
56  $cbr1 set interval_ 0.005
57  $cbr1 attach-agent $udp1
58  $ns at 0.1 "$cbr start"
59  $ns at 0.15 "$cbr1 start"
60  $ns at 2.35 "$cbr stop"
61  $ns at 2.8 "$cbr1 stop"
62  $ns at 3.0 "finish"
63
64  proc  finish {} {
65    global tf nf
66    close $tf
67    close $nf
68    exec nam traceExp.nam
69    exit 0
70  }
71  #Run Simulation
72  $ns run
```

Now it is time to write some AWK scripts to extract data from the trace file. Listing 6.23 extracts the number of packets received, the number of packets dropped, and the average delay incurred by each packet. Lines 5 to 15 are used to initialize different variables. In lines 17 to 23, different column values of a scanned event trace lines are stored in different variables. Lines 27 to 29 are used to record the packet sending times, and lines 31 to 34 are used to record the packet arrival times at the receiving end. The sending time is considered when the packet is enqueued at the source (+ event); columns 3 and 4 in a event trace line represent the intermediate source and destination, whereas columns 9 and 10 represent the original source and destination to which this packet belongs for any event. When a packet travels from node 0 to node 5 (*flow_*1), it actually moves from 0-2, 2-3, and 3-5. Each time it passes through a different link, the enqueue and dequeue operation is performed (three times in this case). In other words, one packet undergoes three sendings and three receivings. But to calculate the delay, we need to consider the times at which the first sending and last receiving occur. To do so, the conditions *intSrc* $==$ *src* and *intDest* $==$ *dest* are used in lines 27 and 31, respectively. The *intSrc, intDest* variables represent the intermediate source and destination, whereas *src, dest* represent the actual source and destination. Lines 35 to 37 are used to count the number of packets dropped.

As usual, lines 16 to 37 are executed for each line present in the input trace file (traceExp.tr) and it moves to line 39 when all the lines are scanned and processed. By this time, all the sending and arrival times are recorded and stored in two arrays, *pktSent* and *pktRecvd*. In line 45, individual delays are calculated and summed; finally, the average delay is determined in line 49. At the end, it prints all the values.

Listing 6.23: Script to find average delay

```
1  # ====== avgDelay.awk ======
2  #!/bin/awk -f
3  # Script to find out average delay in a wired network.
4  # Run with command: awk -f avgDelay.awk < traceExp.tr
5  BEGIN  {
6      for  (i in pktSent) {
7          pktSent[i] = 0
8      }
9      for (i in pktRecvd) {
10          pktRecvd[i] = 0
11      }
12      totPktRecvd=0
13      delay = avgDelay = 0
14      drop=0
15  }
16  {
17      event = $1
18      time = $2
19      intSrc=$3
20      intDest=$4
21      src=$9
22      dest=$10
23      pktSeq = $12
24      intSrc = (intSrc".0")
25      intDest = (intDest".0")
26      # Store packets send times
27      if (pktSent[pktSeq] == 0 &&  event == "+" && intSrc == src) {
28          pktSent[pktSeq] = time
29      }
30      # Store packets arrival time
31      if (event == "r" && intDest == dest) {
32          totPktRecvd ++
33          pktRecvd[pktSeq] = time
34      }
35      if (event == "d") {
36          drop++
37      }
38  }
39  END {
40      # Compute average delay
41      for (i in pktRecvd) {
42      if  (pktSent[i] == 0) {
43              printf("\nError %g\n",i)
44          }
45          delay += pktRecvd[i] - pktSent[i]
46          num ++
```

```
47          }
48          if (num != 0) {
49              avgDelay = delay / num
50          } else {
51              avgDelay = 0
52          }
53          printf("Total Packets Recvd = %g \n",totPktRecvd)
54          printf("Total Packets Dropped = %g \n",drop)
55          printf("Avg Delay = %g ms\n",avgDelay*1000)
56  }
```

The output of the above program may be similar to the following:

```
Total Packets Recvd = 672
Total Packets Dropped = 245
Avg Delay = 89.3015 ms
```

The AWK script in Listing 6.24 does the same thing, but without calculating the average delay. It determines the delay incurred by *flow_1* only and records the sending and receiving times. Finally, it prints the information about the first 100 packets along with dropped packets and time of drop. The delay information is redirected to a file called *delay.dat* in line 44. This file contains three columns: the first column is the packet number, the second is the packet sending time, and the third column represents the packet receiving time. Similarly, it redirects the packet number and drop time information to another file called *drop.dat* in line 49. These values may now be plotted in a graph to visualize the variation in packet delay.

Listing 6.24: AWK script to find packet wise delay and packet drop

```
1   # ====== delayVariation.awk ======
2   #!/bin/awk -f
3   # Script to find out packet wise delay of one flow in a wired
        network.
4   # Run with command: awk -f delayVariation.awk < traceExp.tr
5   BEGIN {
6     for (i in pktSent) {
7       pktSent[i] = 0
8     }
9     for (i in pktRecvd) {
10        pktRecvd[i] = 0
11    }
12    totPktRecvd=0
13    delay = avgDelay = 0
14    flow = 1
15  }
16  {
17    event = $1
18    time = $2
19    intSrc=$3
20    intDest=$4
21    fid=$8
22    src=$9
23    dest=$10
```

```
24   pktSeq = $11
25   intSrc = (intSrc".0")
26   intDest = (intDest".0")
27   # Store packet sending times
28       if (pktSent[pktSeq] == 0 &&  event == "+" && intSrc == src
         && fid == flow) {
29     pktSent[pktSeq] = time
30   }
31   # Store packets arrival time`
32   if (event == "r" && intDest == dest && fid == flow) {
33     totPktRecvd ++
34       pktRecvd[pktSeq] = time
35   }
36   if (event == "d" && fid == flow) {
37       drop++
38       pktDrop[pktSeq]= time
39   }
40 }
41 END {
42   for (j=1;j<=375;j++) {
43     if ( pktRecvd[j] > 0) {
44           printf("%g\t%g\t%g\n", j,pktSent[j], pktRecvd[j]) >
       "delay.dat"
45           }
46     }
47     for (j=1;j<=100;j++) {
48         if (j in pktDrop){
49             printf("%g\t%g\n",j,pktDrop[j]) > "drop.dat"
50         }
51     }
52 }
```

The Gnuplot script in Listing 6.25 plots the delay variation and packet drop information in one plot for the first 100 packets of $flow\_1$ from the data files generated by AWK script in Listing 6.24.

Listing 6.25: Gnuplot script to plot delay variation

```
1 # ======= delayVariation.p =================
2 # plots packetwise delay variation
3 # run command: gnuplot> load 'delayVariation.p'
4       set autoscale
5       unset log
6       unset label
7       set xtic auto
8       set ytic auto
9       set xlabel "time (secs)"
10      set ylabel "packet#"
11      set title "Packet wise delay variation with packet drop
      of flow_1"
12      set key right bottom #legend
13      set grid
14      plot "delay.dat" using 2:1 title 'Packet Sent' with
      lines lw 2, "delay.dat" using 3:1 title 'Packet Received'
      with lines lw 2,  "drop.dat" using 2:1 title 'Packet drop'
      with p pt 7
```

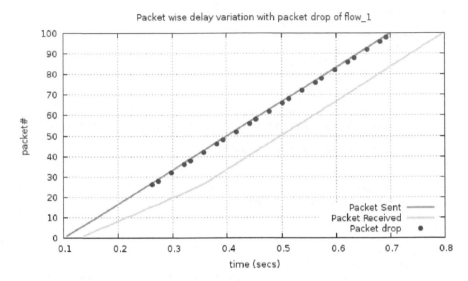

Figure 6.32: Delay variation

The resulting plot is provided in Figure 6.32. It can be seen that the packet sending rate remains constant as generated by CBR, for which it plots a perfectly straight line. But the packet receiving time is initially low, as there is only one flow active. When the *flow_2* starts, the packet delay for *flow_1* increases due to queueing delay. Packet drops are shown as dots. Dropping of packets starts just before 0.3 sec, as the queue cannot accommodate more packets.

The AWK script in Listing 6.26 may be used to find the average throughput separately for *flow_1*, *flow_2* and for the network. Lines 5 to 11 are used to initialize the variables, and lines 13 to 20 are used to store different values from the trace in different variables. Lines 22 to 25 calculate total bytes received for *flow_1* and record the total time of communication. Similarly, lines 27 to 30 find the total bytes received by *flow_2* as well as the simulation time.

Listing 6.26: AWK script to find average throughput

```
1  # ========= avgThput.awk ==========
2  #!/bin/awk -f
3  # Script to find out average throughput of a wired network.
4  # Run with command: awk -f avgThput.awk < traceExp.tr
5  BEGIN  {
6      thPut1 = 0
7      thPut2 = 0
8      thPut = 0
9      simTime1 = 0
10     simTime2 = 0
11 }
```

```
12  {
13        event = $1
14     time = $2
15     intSrc = $3
16     intDest = $4
17     src = $9
18     dest = $10
19     size = $6
20     fid = $8
21     # Find total bytes sent by flow_1
22     if (event == "r" && fid == 1 && intDest == dest) {
23       thPut1 += size
24         simTime1 = time
25         }
26     # Find total bytes sent by flow_2
27     if (event == "r" && fid == 2 && intDest == dest) {
28       thPut2 += size
29         simTime2 = time
30         }
31  }
32  END {
33    thPut = thPut1 + thPut2
34    thPut1 = (thPut1 * 8)/(1000000 * simTime1 )
35    thPut2 = (thPut2 * 8)/(1000000 * simTime2 )
36    if(simTime1<simTime2) {
37        simTime = simTime2
38    } else {
39        simTime = simTime1
40    }
41        thPut = (thPut * 8)/(1000000 * simTime )
42    printf("Avg throughput of flow_1 = %g Mbps\n", thPut1)
43    printf("Avg throughput of flow_2 = %g Mbps\n", thPut2)
44    printf("Network throughput = %g Mbps\n",thPut)
45  }
```

Network throughput is nothing but the throughput achieved by all flows in a network, calculated in line 33. Lines 34 to 41 are used to convert total bytes into throughput in Mbps for different flows. Lines 42 to 44 print different throughput as follows:

```
Avg throughput of flow_1 = 0.301333
Avg throughput of flow_2 = 0.594667
Network throughput = 0.896
```

The script in Listing 6.27 provides the variation in throughput for both the flows and for the network with respect to time. Each time a packet is received, the throughput is calculated and recorded along with the time. This process is repeated for $flow\_1$ in lines 29 to 32 and for $flow\_2$ in lines 34 to 37. Finally, in lines 44 to 61, all the above recorded values are redirected to three different files.

Listing 6.27: AWK script to find throughput variation

```
1  # ====== thputVariation.awk ======
```

```awk
 2  #!/bin/awk -f
 3  # Script to find out average throughput of a wired network.
 4  # Run with command: awk -f thputVariation.awk < traceExp.tr
 5  BEGIN  {
 6      for (i in thPut1){
 7      thPut1[i] = -1
 8      }
 9    for (i in thPut2){
10      thPut2[i] = -1
11    }
12    for (i in thPut){
13      thPut2[i] = -1
14    }
15    thput = 0
16    thput1 = 0
17    thput2 = 0
18  }
19  {
20    event = $1
21    time = $2
22    intSrc = $3
23    intDest = $4
24    src = $9
25    dest = $10
26    size = $6
27    fid = $8
28    # Find throughput for flow_1
29      if (event == "r" && fid == 1 && intDest == dest) {
30      thput1 += size
31        thPut1[time] = (thput1 * 8)/(1000000 * time )
32    }
33    # Find throughput for flow_2
34    if (event == "r" && fid == 2 && intDest == dest) {
35      thput2 += size
36      thPut2[time] = (thput2 * 8)/(1000000 * time )
37    }
38    # Find Network throughput
39    if (event == "r" && intDest == dest) {
40      thput += size
41      thPut[time] = (thput * 8)/(1000000 * time )
42    }
43  }
44  END {
45    for (i in thPut1){
46      if (thPut1[i]>=0){
47        printf("%g\t%g\n",i,thPut1[i]) > "thput1.dat"
48      }
49    }
50    print "\n\n"
51    for (i in thPut2){
52      if (thPut2[i]>=0){
53          printf("%g\t%g\n",i,thPut2[i])> "thput2.dat"
54        }
55    }
56    print "\n\n"
57    for (i in thPut){
```

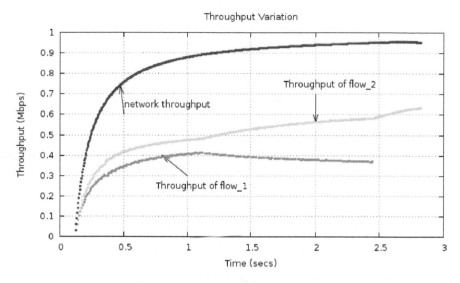

Figure 6.33: Throughput variation

```
58      if (thPut[i]>=0){
59          printf("%g\t%g\n",i,thPut[i])> "thput.dat"
60      }
61  }
62 }
```

A Gnuplot script is written in Listing 6.28 to plot the throughput variation for different flows as generated by the above AWK script. The resultant plot is provided in Figure 6.33, where it can be observed that the network throughput curve follows the ideal throughput curve just below 1 Mbps due to some packet losses (drops). The throughput curve for $flow\_1$ goes below the curve of $flow\_2$ because, more packets belonging to $flow\_1$ are dropped as compared to $flow\_2$. This kind of scenario is common when we use simple queueing mechanisms like the drop tail queue.

Listing 6.28: Gnuplot script to plot throughput variation

```
1 # ======= thputVariation.p ==================
2 # plots variation of throughput w.r.t. time
3 # run command: gnuplot> load 'thputVariation.p'
4 set autoscale
5 unset log
6 unset label
7 set xtic auto
8 set ytic auto
9 set xlabel "Time (secs)"
10 set ylabel "Throughput (Mbps)"
11 set title "Throughput Variation using Fair Queuing"
12 unset key # no legends
```

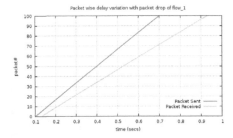
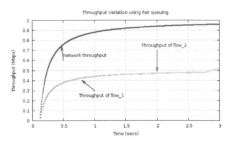

Figure 6.34: Delay variation with fair queueing

Figure 6.35: Throughput variation with fair queueing

```
13 set arrow from 0.5,0.6 to 0.47,0.74
14 set label "Network throughput" at 0.5,0.65
15 set arrow from 2,0.7 to 2.0,0.49
16 set label "Throughput of flow_2" at 1.75,0.75
17 set arrow from 1,0.3 to 0.8,0.4
18 set label "Throughput of flow_1" at 0.75,0.25
19 set grid
20
21 plot "thput1.dat" using 1:2  with p pt 7 ps 0.4, "thput2.dat"
       using 1:2  with p pt 7 ps 0.4, "thput.dat" using 1:2  with p
       pt 7 ps 0.4
```

To avoid uneven packet losses, other mechanisms such as fair queueing may be used. If you change line 25 of Listing 6.22 given above as follows:

```
16. $ns duplex-link $n2 $n3 1Mb 10ms FQ
```

then the queueing policy will be fair, that is, the mechanism will take care to drop approximately the same number of packets from both flows and it tries to minimize packet drops. The resultant graphs are shown in Figures 6.34 and 6.35. Two major points can be noted from these graphs. No packet loss occurs within the first 100 packets, and the throughput of both flows is nearly equal, which essentially is the advantage of using fair queueing. However, the queueing delay is greater compared to the drop tail strategy.

Test/Viva Questions

    a) Consider any event trace file that contains data in various columns. What is the difference between the data in columns 3 and 4 and 9 and 10?

    b) Distinguish between columns 11 and 12 of an event trace format.

    c) What is the significance of column 9 in an event trace format?

    d) Briefly explain whether can you find only queueing delay?

    e) What is the relation between bandwidth and throughput?

    f) Prepare a formula to find the percentage packet loss in a simulation.

    g) What is the difference between packet rate and packet interval?

Programming Assignments

1) Write AWK script to find the average delay for separate flows from the trace file traceExp.tr, then write a Gnuplot script to plot a histogram/bar chart to show individual average delay along with combined averaged delay for both drop tail and fair queueing.

2) Find the average throughput for different flows by varying the CBR packet size from 200 to 1000 bytes, with a step size of 100 bytes. Now plot these values in a line graph.

3) Find the average throughput for different flows by varying the CBR packet interval from 0.001 to 0.006, with a step size of 0.001. Now plot these values in a line graph.

4) Modify Listing 6.22 to use two ftp traffics on the top of TCP. Now extract and plot delay and throughput variations for different flows.

5) Modify Listing 6.22 to use one ftp and one CBR traffic. Extract and plot delay and throughput variations for different flows using drop tail, FQ, and SFQ queueing models.

6.8 Application Layer — Traffic Generators

The application layer is the top-most layer in any network model, and it sits on the top of the transport layer. In NS, there are two types of applications: *traffic generators* and *simulated applications*. *Traffic generators* are used at the top of UDP agents, and *simulated applications* are used at the top of TCP agents. The different traffic generators and simulated applications currently available in NS are shown in Figure 6.36.

Commands

```
set ftp1 [new Application/ftp]
$ftp1 attach-agent $src
```

The following shortcut accomplishes the same result.

```
set ftp1 [$src attach-app ftp]
```

6.8.1 Traffic generators

There are four types of traffic generators supported in NS as follows.

i) **Exponential Traffic**
Generates traffic according to exponential on/off distribution.
on state: packets are generated/sent at a constant burst rate
off state: no traffic is generated/sent
where *burst time* and *idle times* are taken from exponential distributions.
Configuration parameters are

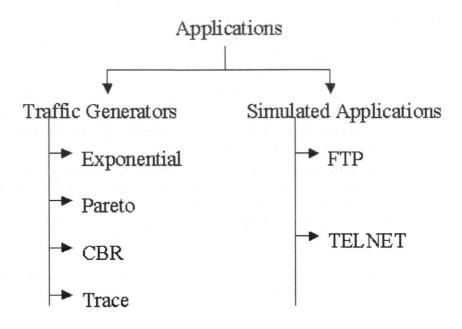

Figure 6.36: Applications/traffic generators in NS

> *packetSize_*   ;# generated packet size
>
> *burst_time_*   ;#the average "on" period
>
> *idle_time_*   ;#the average "off" period
>
> *rate_*   ;#packet sending rate during "on" times

Example:

```
set exp1 [new Application/Traffic/Exponential]
$exp1 set packetSize_ 210
$exp1 set burst_time_ 500ms
$exp1 set idle_time_ 500ms
$exp1 set rate_ 100k
```

Note: When *burst_time_* is set to 0 and *rate_* to a very large value, the *exponential* generator behaves as a *Poisson* generator.

ii) **Pareto Traffic**

Generates on/off traffic according to pareto on/off distribution. Configuration parameters:

> *packetSize_*
>
> *burst_time_*

*idle_time_*

*rate_*

*shape_* ;# used by the Pareto distribution

Example:

```
set par1 [new Application/Traffic/Pareto]
$par1 set packetSize_ 512
$par1 set burst_time_ 520ms
$par1 set idle_time_ 480ms
$par1 set rate_ 200k
$par1 set shape_ 1.5
```

iii) **CBR Traffic**

Generates packets at a constant bit rate.

Configuration parameters:

*rate_* ;# packet sending rate
*interval_* ;# interval between packets
*packetSize_* ;# size of the packets in bytes
*random_* ;# flag to introduce random noise in the scheduled departure times (default is off)
*maxpkts_* ;# maximum number of packets to send (default is 228)

Note: The settings of *rate_* and *interval_* are mutually exclusive, that is, either a *rate_* or an *interval_* has to be configured in a simulation, but both of them cannot be used simultaneously.

Example:

```
set cbr1 [new Application/Traffic/CBR]
$cbr1 set packetSize_ 48
$cbr1 set rate_ 64Kb
$cbr1 set random_ 1
```

iv) **Trace**

In this case the traffic is generated according to an input *trace* file. Each record in the trace file contains two fields. The first field represents *time* in μsec until the last packet was generated, and the second field represents the length of the packet in bytes. There are no configuration parameters and no restriction to the number of records in the file.

Example:

```
set tFile [new Tracefile]
$tFile file-name ex-trace
```

```
set t1 [new Application/Traffic/Trace]

$t1 attach-tracefile $tFile

set t2 [new Application/Traffic/Trace]

$t2 attach-tracefile $tFile
```

Random starting times are to be chosen for different trace objects (t1, t2) within the trace in order to avoid synchronization of the traffic.

6.8.2 Simulated applications

There are two types of simulated applications supported in NS2 as follows.

i) **ftp**
   Ftp simulates bulk data transfer.
   Commands:
   $ftp start
   $ftp stop
   $ftp attach-agent
   $srcNode
   $ftp produce <n>   ;# Ftp to produce n packets instantaneously

   $ftp producemore   ;# Ftp to produce count more pack-
   <count>   ets

   Configuration Parameter:
   maxpkts_   ;# The maximum number of pack-
                                 ets generated by the source

ii) **Telnet**
   Listing 6.29 demonstrates the use of ftp application on the top of fullTCP. Traffic (packets) are generated in Telnet in either of the two ways given below. Packet intergeneration time is chosen

   i) from an exponential distribution with the average interval as *interval_* for *interval_* $\neq 0$

   ii) from tcplib distribution for *interval_* $= 0$

   Commands:

   $telnet start

   $telnet stop

   $telnet attach-agent $srcNode
   Configuration Parameter:
   interval_   ;# Average packet inter-generation
                                 time in seconds.

Listing 6.29: Ftp application on top of a FullTCP

```
1  set src [new Agent/TCP/FullTcp]
2  set sink [new Agent/TCP/FullTcp]
3  $ns_ attach-agent $node_(s1) $src
4  $ns_ attach-agent $node_(k1) $sink
5  $ns_ connect $src $sink
6  # set up TCP-level connections
7  $sink listen;
8  $src set window_ 100
9  set ftp1 [new Application/FTP]
10 $ftp1 attach-agent $src
11 $ns_ at 0.0 "$ftp1 start"
12 $ns_ at 5.0 "$ftp1 stop"
13 $ns_ run
```

Test/Viva Questions

a) Differentiate between *traffic generators* and *simulated applications*.

b) What are the different traffic generators available in NS?

c) Distinguish between exponential and Pareto traffic generators.

d) Explain the *rate_* and *interval_* parameters of CBR traffic.

e) What is the command to create an exponential traffic?

f) How can one to configure exponential traffic to behave as a Poisson process?

g) When does one to use a trace traffic?

h) Distinguish between ftp and Telnet applications.

i) What is the configuration parameter for a Telnet application?

j) What is the configuration parameter for an FTP application?

Programming Assignments

1) Create a network as in Figure 6.27. Attach two UDP agents to node 0 and node 1. Node 0 uses an exponential traffic and node 1 uses the Pareto traffic to send packets to node 5 and node 4, respectively. Configure the values for packet size and burst time same for both traffics. Now measure the normalized throughput of different connections.

2) In a network as in Figure 6.27, Attach one FTP and another Telnet application on top of TCP agents to node 0 and node 1. Node 0 sends to node 5 and node 1 sends to node 4. Now measure the normalized throughput of different connections. Also measure the average delay for both connections.

6.9　Network Dynamics—Node/Link Failure Models

To make the topologies dynamic, NS2 has the provision of introducing link/node failures in a simulation environment.

Command:

`$ns rtmodel <model> <model-params> <args>`

rtmodel provides a failure model to be applied either to the nodes or to the links available in a simulation topology. That is, if the last argument contains two nodes, then it is understood as a link failure between these two nodes. Otherwise if only one node is given in the argument, it is considered to be a failure of that node. Model parameters are different for each model, and *args* specifies the nodes or links to which the failure model will be applied. The various failure models are as follows.

i) **Deterministic Model**

This is an on/off model that takes four parameters as follows.

`<start-time>　<up-interval>　<down-interval>`
`<finish-time>`

Starting from *start-time* (simulation starting time) the link remains active (up) for a duration of *up-interval* and inactive (down) for a period of *down-interval*. This up-down cycle continues until *finish-time* (end of simulation). The default values for model parameters are

start-time: *0.5sec*
finish-time: *end of simulation*
up-interval: *2.0sec*
down-interval: *1.0sec*
Example:

`$ns rtmodel Deterministic 20.0 20.0 $n1 $n5`

That is, starting from 20 s of simulation, the link remains active for 20 s and inactive for the default time. This active/inactive cycle continues until the end of the simulation.

ii) **Exponential Model**

This is also an on/off model that takes four parameters as follows.

`<start-time>　<up-interval>　<down-interval>`
`<finish-time>`

The values of *up-interval* and *down-interval* are considered as the mean of the exponential distribution which defines the time of the up/down cycle for the link/node.

The default values for different parameters are as follows.

start-time: *0.5s*
finish-time: *end of simulation*
up-interval: 10s
down-interval: 1s

To assume a default value, a "-" symbol is used in place of an actual value (see example below).

Example:

0.8 1.0 1.0   ;# start at 0.8s, up/down = 1.0s, finish is default

5.0 0.5   ;# start is default, up/down = 5.0s, 0.5s, finish is default

− 0.7   ;# start, up interval are default, down = 0.7s, finish is default

− − − 10   ;# start, up, down are default, finish at 10s

Example:

A node failure cycle

```
$ns rtmodel Exponential 0.7 1.5 1.0 $n1
```

iii) **Trace Model**

In this model, the link/node dynamics is taken from a trace file. The trace-driven model takes one parameter: the name of the trace file.

```
$ns rtmodel Trace <trace-file>
```

The format of the input trace file is as follows.

```
v <time> link-<operation>  <node1>  <node2>
```

```
v 0.5 link-up 1 2
```

```
v 2.8 link-down 7 3
```

If a line specified in a trace file does not correspond to any node or link, then it is ignored.

iv) **Manual Model**

This model takes two parameters:

operation to be performed

time at which the operation is to be performed.

The manual distribution could be specified alternately using the *rtmodel-at* method as follows.

```
$ns rtmodel-at 3.5 up
```

```
$n0 $ns rtmodel-at 3.9 up $n(3) $n(5)
```

```
$ns rtmodel-at 40 down $n4
```

6.10   Error Model

Up to this point, we have considered that packet losses in a transmission occur due to the buffering limit, link failures, and node failures. But in a real network implementation, packet loss may occur due to various different reasons, like a noisy link, etc. The error models available in NS help us to incorporate random packet losses in a simulation. In general, packet losses normally are more prominent in wireless networks compared to wired networks. The error models in NS may be used for both wired and wireless networks.

Many error models are implemented in NS; some major types are given as follows.

i. **ErrorModel**: Uniform error model

ii. **ErrorModel/TwoState**: Maintains two states, either error free or error. In an error-free state, no packet loss occurs and in a error state, all packets are lost.
Also *ErrorModel/Expo, ErrorModel/Empirical, and ErrorModel/TwoStateMarkov* models are inherited from the above two-state model.

iii. **ErrorModel/List**: Drops packets/bytes in an order specified in the list.

Configuration Parameters:
The following configuration parameters may be used to customize the packet drops in a simulation.

a) **enabled_**: Set to 1, if active, otherwise 0.

b) **rate_**: Probability of error.

c) **delay_pkt_**: If set, error model will not drop the packet but will delay packet transmission.

d) **delay**: Defines time of delay, if above parameter is set.

e) **markecn_**: If set, the model will not drop the packet but marks an error flag in the packet header.

f) **unit**: Defines loss unit. It can be any one from {pkt, byte, bit, and time}.

g) **ranvar**: Random variable type.

. . .

Let us consider a simple two-node network, and attach an error model as given in Listing 6.30. Lines 1 to 12 as usual deal with creating nodes, links, and traces. Lines 15 to 20 are extra to insert an error model into a simulation. In line 15, a simple uniform error model is chosen with a probability of 0.15 in line 16. That is, the model will try to drop 15% of packets (line 16), and how to chose these packets is decided by the ranvar parameter in line 17. Where to drop these packets is given in line 18, that is, at the destination. Finally, line 19 tells the simulation to attach the error to a link. The remaining lines are used to attach an UDP agent, CBR traffic to the simulation, and the traffic start/stop times.

When this program is executed, it runs the animation, and it can be seen that some of the packets are dropped at n2, the receiving end.

Listing 6.30: Uniform error model in a two-node network

```
1  set ns [new Simulator]
2
3  #trace files
4  set tf [open errorExp.tr w]
5  $ns trace-all $tf
6  set nf [open errorExp.nam w]
7  $ns namtrace-all $nf
8
9  #two nodes and a link between them
10 set n1 [$ns node]
11 set n2 [$ns node]
12 $ns duplex-link $n1 $n2 10Mb 1ms DropTail
13
14 #configure error model
15 set em [new ErrorModel]
16 $em set rate_ 0.15
17 $em unit pkt
18 $em ranvar [new RandomVariable/Uniform]
19 $em drop-target [new Agent/Null]
20 $ns link-lossmodel $em $n1 $n2
21
22 #CBR at the top of UDP
23 set udp [new Agent/UDP]
24 set null [new Agent/Null]
25 set cbr [new Application/Traffic/CBR]
26 $ns attach-agent $n1 $udp
27 $ns attach-agent $n2 $null
28 $cbr attach-agent $udp
29 $ns connect $udp $null
30
31 $ns at 1.0 "$cbr start"
32 $ns at 10.0 "$cbr stop"
33 $ns at 10.000001 "finish"
34
35 proc  finish {} {
36            global tf nf
37            close $tf
38            close $nf
39            exec nam errorExp.nam &
40            exit 0
41       }
42
43 #run simulation
44 $ns run
```

To know the exact count of packets dropped and received, the following AWK script may be used.

Listing 6.31: AWK script to find percentage packet drop

```
1  # ======= drop.awk =======
2  # run: awk -f drop.awk < errorExp.tr
3
4  BEGIN {
5    totPktRecvd=0
6    dropPkts=0
```

```
7  }
8  {
9    event = $1
10   if (event == "r" ) {
11     totPktRecvd ++
12       }
13         if (event == "d") {
14     dropPkts++
15   }
16 }
17 END {
18   printf("Total Packets Recvd = %g \n",totPktRecvd)
19         printf("Total Packets Dropped = %g \n",dropPkts)
20         printf("Pkt drop rate= %g \n",dropPkts/(dropPkts+
     totPktRecvd))
21 }
```

The packet drop rate that we get from the above program may not converge to the rate of packet drop defined in line 16 of Listing 6.30. To test the packet drop rate for different simulation intervals, we need to run the simulation experiments multiple times. However, we can write a shell script, as given in Listing 6.32, that runs the simulation multiple times automatically with different simulation intervals for each run. As can be seen in line 7 of Listing 6.32, the simulation is executed with two command line arguments passed to the simulation. The first argument is the simulation time, and the second argument is the coresponding trace file name, as we need different trace files for different output. Again the rate found from each run may be extracted from the corresponding trace file by calling an AWK script at line 8 that stores the simulation interval times and actual packet drop rates as two columns in a text file. As this results in very large trace files, after data extraction, the trace files may be deleted, as in line 9. But if you want to keep the trace files, then just comment line 9.

Listing 6.32: Shell script to run a simulation multiple times

```
1  #! /bin/bash
2  #script to run errorModel.tcl multiple times
3  #each time with a different simulation time
4  # run: bash shellscript.sh
5
6  simScript="./errorModel.tcl"
7  let limit=2500
8  let step=100
9
10 for(( simTime=100; simTime<=limit; simTime+=step ))
11   do
12   echo "Simulating for Sim Time=$simTime ...."
13   #run simulation
14   ns $simScript -time $simTime -tr ./errorExp_$simTime.tr
15   #Find drop statistics
16   awk -f dropStat.awk < ./errorExp_$simTime.tr
17   #delete trace file
18   rm ./errorExp_$simTime.tr
```

```
19  done
```

Listing 6.33 gives the simulation program that takes two command line arguments as supplied by the shell script given in Listing 6.32. It uses two extra procedures. The procedure *usage* is used to handle the situation if the required number of arguments is not supplied from the command line. The procedure *getopt* is used to extract the command line argument values and stores these in an associative array *opt*. The remaining lines in the program are self-explanatory.

Listing 6.33: Error model program with command line arguments

```
1  # errorModel.tcl
2  # Program that takes command line arguments
3  # it is executed by shellscript.sh
4
5  set opt(time)      ""
6  set opt(tr)        ""
7  proc  usage {} {
8    global argv0
9    puts "\nusage:\n ns $argv0  \[-time simTime \]  \[-tr trace
       file \]\n"
10 }
11
12 proc getopt {argc argv} {
13   global opt
14     for  {set i 0} {$i < $argc} {incr i} {
15     set arg [lindex $argv $i]
16     if {[string range $arg 0 0] != "-"} continue
17         set name [string range $arg 1 end]
18         set opt($name) [lindex $argv [expr $i+1]]
19     }
20 }
21
22 # get commandline options
23 getopt $argc $argv
24 if { $opt(time) == "" || $opt(tr) == "" } {
25   usage
26   exit
27 }
28
29 # create simulator instance
30 set ns [new Simulator]
31
32 # create trace object for ns and nam
33 set tf [open $opt(tr) w]
34 $ns trace-all $tf
35
36 set n1 [$ns node]
37 set n2 [$ns node]
38
39 $ns duplex-link $n1 $n2 10Mb 1ms DropTail
40
41 set em [new ErrorModel]
42 $em set rate_ 0.15
```

```
43 $em unit pkt
44 $em ranvar [new RandomVariable/Uniform]
45 $em drop-target [new Agent/Null]
46 $ns link-lossmodel $em $n1 $n2
47
48 set udp [new Agent/UDP]
49 set null [new Agent/Null]
50 set cbr [new Application/Traffic/CBR]
51 $ns attach-agent $n1 $udp
52 $ns attach-agent $n2 $null
53 $cbr attach-agent $udp
54 $ns connect $udp $null
55
56 $ns at 1.0 "$cbr start"
57 $ns at $opt(time) "$cbr stop"
58 $ns at $opt(time).000000002 "puts \"Simulation done.\" ; $ns
      halt"
59
60 #start simulation
61 puts "Starting Simulation..."
62 $ns run
```

Listing 6.34 is same as Listing 6.31 except for the print statements in line 19. Here we print two values only in two columns of a text file called drop-Stat.dat, so that we can plot a graph from that data file.

Listing 6.34: AWK script to extract dropped packet percentage

```
1  # ====== dropStat.awk ======
2  # used in shellscript.sh to find packet drop rate
3
4  BEGIN  {
5     totPktRecvd=0
6     dropedPkts=0
7     time=0
8  }
9  {
10    event = $1
11    t=$2
12    if  (event == "r" ) {
13      totPktRecvd ++
14        time=t
15    }
16    if (event == "d") {
17      dropedPkts++
18      time=t
19    }
20  }
21
22  END {
23    printf(" %d\t %g \n",time+1, dropedPkts/(dropedPkts +
      totPktRecvd)) >> "dropStat.dat"
24
25  }
```

Now use the Gnuplot script given in Listing 6.35 to plot the convergence

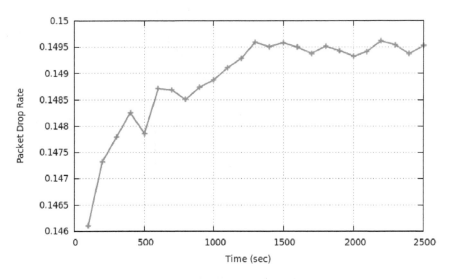

Figure 6.37: Packet drop rate vs. simulation time interval

rate of error probability with respect to simulation time. The output plot is provided in Figure 6.37.

Listing 6.35: Gnuplot script to plot dropped packet percentage

```
1  #------plotDropStat.p------
2  # Plots packet drop rate of different simiulations obtained
       running shellscript.sh
3  set autoscale
4  unset log
5  unset label
6  set xtic auto
7  set ytic auto
8  set xlabel "Simulation Time (sec)"
9  set ylabel "% Packet Drop Rate"
10 set title "% "
11 unset key #no legend
12 set grid
13 plot "dropStat.dat" using 1:2  with lp lw 2
```

As can be seen from the figure, the convergence rate increases as the simulation time increase. From 1100 sec onwards, it remains closer to the defined value.

Test/Viva Questions

a) Distinguish between network dynamics and error model in NS.

b) What is the difference between exponential and deterministic models of link failures?

c) How can a node failure be programmed in a simulation script?

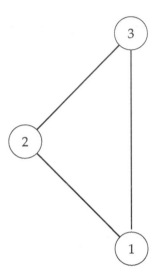

Figure 6.38: Three-node network

d) The arguments to an exponential model of failure are 0.5 1.5 - 3. Explain the meaning of these values.

e) What is the command to manually cause failure of a node at a specific time?

f) What are the different error models available in NS2?

g) What is the use of the markecn_ parameter in an error model?

h) What is the use of the delay_pkt_ parameter in an error model? What other parameters are required to be set for this parameter?

i) What are different loss units?

Programming Assignments

1) Design a three-node network as given in Figure 6.38. Attach a CBR traffic to the UDP connection between node 1 and node 3. Attach a two-state error model with 90% good and 10% bad periods. Find the packet drop rate with varying simulation time intervals.

2) Design a three-node network as given in Figure 6.38. Attach a CBR traffic to the UDP connection between node 1 and node 3. The link 1–3 fails using any deterministic model and it uses some dynamic routing, that is, it takes on the alternate path for packet forwarding. Attach a uniform error model with the probability of packet drop set as 0.01. Plot the variation of packet drop rates with varying simulation time.

Bibliography

[1] NS Manual (http://www.isi.edu/nsnam/ns/ns-documentation.html)

[2] Marc Greies, Tutorial for the network simulator NS: http://www.isi.edu/nsnam/ns/tutorial/

[3] Altman and Jimenez, NS for beginners: http://www-sop.inria.fr/members/Eitan.Altman/COURS-NS/n3.pdf

[4] Jae Chung and Mark Claypool, NS by Example, http://nile.wpi.edu/NS/Advanced Reading

CHAPTER 7

WIRELESS NETWORK SIMULATION

Wireless networks may be divided into fixed wireless networks and mobile wireless networks. In the first category, nodes are stationary in the sense that the movement of nodes is restricted within the coverage area of the network equipment. A prominent example of a stationary wireless network is a wireless LAN. While in a mobile wireless network, the nodes can change their positions with respect to time and move from the coverage area of one set of network equipment to another. A mobile wireless network can further be classified into two types. A mobile wireless network in which no fixed networking devices are used (such as routers, switches, base stations, access points, etc.) is called a *mobile wireless ad hoc network* or *infrastructureless network*. On the other hand, in an infrastructure-based network, networking components are fixed with respect to time. In this chapter, we discuss simulation experiments for various types of wireless networks.

The wireless extension to NS2 was originally contributed by Carnegie Melon University's (CMU) *Monarch Project*. Different modules were attached to NS2 to simulate wireless networks for this purpose along with modules to handle node mobility. The basic differences between a wired network node and a wireless node are essentially the following

Mobile Node: A mobile node incorporates functionalities for supporting wireless communications. These include the capability to receive and/or transmit signals through a wireless channel and the ability to move within a given network boundary. Unlike a wired network node, a mobile node is not connected physically by means of links to other fixed or mobile nodes.

Routing: The packet routing process in a wireless network is in many ways fundamentally different from that of the wired network. Routing in a wired network deals with a fixed topology. In the case of

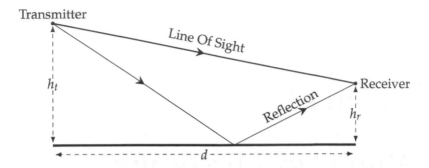

Figure 7.1: Two-ray ground propagation

wireless networks the topology frequently changes and link failure is very common. Therefore the routing protocol is required to be intelligent enough to handle these situations.

Medium Access Control: In a wireless network, MAC protocols such as CS-MA/CD do not work satisfactorily due to *hidden and exposed terminal problems.* Also, a node can not hear collisions since the signal from the collision is extremely faint compared to the transmitted signal. Therefore IEEE 802.11 is used as the MAC sublayer protocol.

Radio Propagation Model: Wireless communication is done via radio waves. Two popular radio propagation models are the following.

   i) ***Free-Space Model***
   This is a simple propagation model in which the communication range is represented in a circular area centered at the transmitter. Communication is possible if a receiver is present within that circle. This model of propagation assumes that no obstacles are present between a sender and a receiver, implicitly thereby assuming a clear line-of-sight path, as shown by a straight line in Figure 7.1. Consequently, the following expression can be used to compute the received signal power per unit area.

$$P_r(d) = \frac{P_t G_t G_r \lambda^2}{(4\pi)^2 d^2 L}$$

where
d: distance of receiver from the sender
P_t: transmitted signal power
G_t and G_r: antenna gains of the sender and receiver, respectively
$L(L \geq 1)$: system loss
λ: wavelength

Normally, the values of G_t, G_r, and L are chosen to be 1 in NS simulation experiments. This model is suitable when the communication range is moderate with only line-of-sight propagation. This model is difficult to use when line-of-sight communication cannot take place due to obstacles.

ii) *Two-Ray Ground*

This model is a more practical model in which both LOS (line of sight) and ground reflection paths (see Figure 7.1) are considered. This model is normally used in long distance transmission (large d) as it provides more accurate prediction than the free-space model. The received power may be computed using the following formula.

$$P_r(d) = \frac{P_t G_t G_r h_t^2 h_r^2}{d^4 L}$$

where

h_t and h_r: heights of the sender and receiver from the ground
As can be seen from the above equation, the signal attenuation (power loss) is much faster than the previous model.

For short distance communications, the two-ray model provides less accurate results than the free-space model. Since two different models are used for two distance ranges, the distance that is used to decide the model is called the *cross-over distance* (d_c).

Let us consider a cross-over distance d_c where,
$d > d_c$: the *two-ray ground* model is used
$d < d_c$: the *free-space* model is used
Therefore, when $d = d_c$, both models produce identical results.

$$\frac{P_t G_t G_r \lambda^2}{(4\pi)^2 d_c^2 L} = \frac{P_t G_t G_r h_t^2 h_r^2}{d_c^4 L}$$

from which the value of the cross-over distance can be found to be

$$d_c = (4\pi h_t h_r) / \lambda$$

7.1 Wired versus Wireless Network Simulation

Wireless network simulation, unlike wired network simulation, may perform the following extra activities.

i. A two-dimensional area (flat space) of some fixed size is considered. This a test bed/grid within which all the nodes are to be placed. For

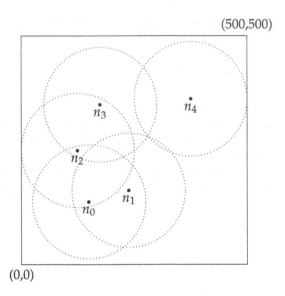

Figure 7.2: Wireless test bed with node communication ranges

example, if the area defined is 500×500, that defines a test bed of size 500 meter $\times$ 500 meter. All the nodes need to be placed within this area.

ii. The nodes used here are mobile nodes, which have different character-istics from wired network nodes. Each node has a circular disk centered on itself, called the communication range of the node, shown as dotted circles Figure 7.2. Mobile node configuration is discussed in more detail in the next subsection.

iii. No links need to be created explicitly, as air is used as the broadcast medium. That is, a node n0 can communicate with node n1, if the node n1 either comes directly in the communication range of node n0 or there exists a third node n2 that is situated in the communication range of both n0 and n1. This situation is shown in Figure 7.2, where n0 can communicate with both n1 and n2, but not with n4. But n0 can com-municate with n3 via n2 as n2 is situated in the communication range of n0 and n3. The node n4 cannot communicate with any other node as no node exists in its communication range.

iv. Nodes can be mobile in a wireless simulation. That is, a node can move from one place to another at any time while the simulation is on.

v. Wireless simulation uses an internal object called GOD (general opera-tions director), that maintains a neighbor list of each node and is used by the simulator internally. Therefore, in each wireless simulation, this object is created.

vi. Some new commands are used for NAM, like *trace, node color, node size, node reset*, etc.

7.2 Step-by-Step Wireless Network Simulation

Let us understand the steps to be followed to simulate a mobile wireless network effectively.

Step I: **Creating the Environment**
instantiate simulator object

```
set ns [new Simulator]
```
#Create event trace

```
set tracefd [open simple.tr w]
```

```
$ns trace-all $tracefd
```
#Create NAM trace

```
set nf [open simple.nam w]
```

```
$ns namtrace-all-wireless $nf 500 500
```
#Create Topography

```
set topo [new Topography]
```

```
$topo load_flatgrid 500 500
```
#Create GOD object #**GOD** is created for internal use of the simulator. It actually stores an array in which it stores the shortest number of hops to reach to each node

```
create-god <number of nodes>
```

Create channel

```
set wChan [new Channel/WirelessChannel]
```

Step II: **Mobile Node Configuration**
There are some parameters that need to be set before the mobile node is created, so that a created node will take on those parameters.

```
$ns node-config
    -adhocRouting <routing Protocol> \
    -macType <MAC Protocol> \
    -ifqType <Interface queue type> \
    -ifqLen <queue length in packets> \
    ...
```

Example:

```
$ns node-config
    # routing protocol is DSR
    -adhocRouting DSR \
    -llType LL\
    #uses MAC 802.11
    -macType Mac/802_11 \
    #Priority queue is used
    -ifqType Queue/DropTail/PriQueue \
    # Max 50 packets in the queue
    -ifqLen 50 \
    ...
```

Step III: Creating Mobile Nodes

To create a node, we need to use the same command as for wired network node creation:

```
set n0 [$ns node]
set node_(1) [$ns node]
```

Set Node size in NAM Display (optional)

```
$ns initial_node_pos $n0 20
```

```
$ns initial_node_pos $n1 30 ;# bigger size node
```
#set node color in NAM display (optional)

```
$n0 color blue
```

```
$ns at 0.0 "$n0 color blue"
```

Step IV: Initializing Node Positions (Optional)

If you are required to place the nodes in specific positions in the flat grid defined, then use the following commands:

```
$node0 set X_ <val1>
$node0 set Y_ <val2>
```
where *val1*, and *val2* are the X and Y coordinate values, respectively, for *node0*. These two lines are repeated with each node available in the network with respective X and Y coordinate values.

Step V: Configuring Node Movements (Optional)

If you need to support node mobility, then use the following syntax to move the nodes from one position to another.
#Syntax to Create Node Movement

```
$node random-motion 0 ;# random motion of nodes is disabled
```

```
$ns at <time> "$node setdest <x> <y> <speed>"
```
node starts moving at time *<time>* sec from its current position to the

new position(<X>,<Y>) in the test bed with a speed of <*speed*> meter/sec. However, before moving the node, the random motion property of the node is disabled by setting it to 0

Example:

\# Node1 starts to move at 50 sec to (25,20) at a speed of 15 m/s.

```
$ns at 50.0 "$node1 setdest 25.0 20.0 15.0"
```

\# Node0 starts to move at 10 sec to (20,18) at a speed of 1.0 m/s.

```
$ns at 10.0 "$node0 setdest 20.0 18.0 1.0"
```

The movements for any node may be attached as many times as desired in a single simulation.

Step VI: **Attaching Transport Agent and Application Traffic**

This step is similar to wired network simulation.

```
set tcp [new Agent/TCP]
set sink [new Agent/TCPSink]
$ns attach-agent $n0 $tcp
$ns attach-agent $n1 $sink
$ns connect $tcp $sink
set ftp [new Application/ftp]
$ftp attach-agent $tcp
$ns at 10.0 "$ftp start"
```

Step VII: **Define Start and Stop times**

\#Set Simulation Stop Time

```
$ns at 150.0 "$node_(0) reset";
```

```
$ns at 150.0 "$node_(1) reset";
```

\# NAM Stop

```
$ns at 150.0 "$ns nam-end-wireless 150.0"
```

```
$ns at 150.0001 "stop"
```

```
$ns at 150.0002 "puts \"NS EXITING...\""
```

\#stop() to be invoked at the end of simulation

```
proc stop {} {
    global ns tracefd nf
    close $tracefd
    close $nf
    exec nam simple.nam &
    exit 0
}
```

Step VIII: **Simulate**

\#To Start The Simulation

```
$ns run
```

Now it is time to write some real simulation programs. Let's first write a very simple program, and then gradually we will add more functionalities to it.

Problem 1: Two Wireless Nodes

Consider two wireless nodes placed in a $500 \times 500m^2$ grid. The node positions are n0 (30,40) and n1 (200,100). An ftp traffic is sent from n0 to n1 over a TCP agent. The complete program is given in Listing 7.1.

Listing 7.1: Simple two-node wireless simulation

```
1  #-----twoNodeWireless.tcl-----
2  #A simple two-node wireless simulation
3
4  #Event scheduler
5  set ns [new Simulator]
6  #trace
7   $ns use-newtrace
8   set tracefd [open wrls.tr w]
9   $ns trace-all $tracefd
10  set namtrace [open wrls.nam w]
11  $ns namtrace-all-wireless $namtrace 500 300
12  # set up topography object
13  set topo [new Topography]
14  $topo load_flatgrid 500 300
15  # Create God
16  create-god 2
17  # Create channel
18  set wChan1 [new Channel/WirelessChannel]
19  # configure a mobile node,
20 $ns node-config -adhocRouting DSDV \
21          -llType  LL \
22      -macType Mac/802_11 \
23      -ifqType Queue/DropTail/PriQueue \
24      -ifqLen 50 \
25      -antType Antenna/OmniAntenna \
26      -propType Propagation/TwoRayGround \
27      -phyType Phy/WirelessPhy \
28      -topoInstance $topo \
29      -agentTrace ON \
30      -routerTrace ON \
31      -macTrace OFF \
32      -movementTrace OFF \
33      -channel $wChan1
34  #create nodes
35  set n0 [$ns node]
36  set n1 [$ns node]
37  #node position
38   $n0 set X_ 30
39   $n0 set Y_ 40
40   $n1 set X_ 200
41   $n1 set Y_ 100
42  #size of nodes in NAM
43   $ns initial_node_pos $n0 20
44   $ns initial_node_pos $n1 30
```

```
45  # Setup traffic flow between nodes
46  set tcp [new Agent/TCP]
47  set sink [new Agent/TCPSink]
48  $ns attach-agent $n0 $tcp
49  $ns attach-agent $n1 $sink
50  $ns connect $tcp $sink
51
52  set ftp [new Application/FTP]
53  $ftp attach-agent $tcp
54  $ns at 3.0 "$ftp start"
55
56  # Tell nodes when the simulation ends
57  $ns at 30.0 "$n0 reset";
58  $ns at 30.0 "$n1 reset";
59  $ns at 30.0 "stop"
60  $ns at 30.01 "puts \"NS EXITING...\" ; $ns halt"
61  proc stop {} {
62      global ns tracefd namtrace
63      $ns flush-trace
64      close $tracefd
65      close $namtrace
66      exec nam wrls.nam &
67      exit 0
68  }
69
70  puts "Starting Simulation..."
71  $ns run
```

The NAM visualization snapshot for above program is given in Figure 7.3 in which the large circles changing continuously represents wireless transmission. These circles are concentric around the nodes. The packet flow are shown as dashed lines moving from node 0 to node 1. The size of nodes are shown to be different due to line 40 and line 41 of the program, where we have defined size 20 for node 0 and size 30 for node 1.

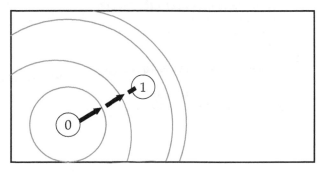

Figure 7.3: Simple two-node wireless simulation

Problem 2: Node Color

In an extension to the above program, let's make the nodes colorful. To do that, add the following lines in Listing 7.1 after the nodes are created in line 33.

```
#node color
$n0 color blue
$n1 color red
$ns at 0.0 "$n0 color blue"
$ns at 0.0 "$n1 color red"
```

When the new program is executed, the node colors in the NAM visualization will be changed to blue and red, respectively, for node n0 and node n1.

Problem 3: Adding Mobility

The nodes can also be mobile. Mobility of the nodes may be added at any time within the simulation time. The complete program is given in Listing 7.2, where lines 34 and 35 and 56 to 61 are added to provide node movement.

Listing 7.2: Two-node wireless simulation with node mobility

```
1  #---twoNodeWireless1.tcl
2  # simulates two node wireless network with node mobility
3  #Event scheduler
4   set ns [new Simulator]
5  #trace
6   $ns use-newtrace
7   set tracefd [open wrls.tr w]
8   $ns trace-all $tracefd
9   set namtrace [open wrls.nam w]
10  $ns namtrace-all-wireless $namtrace 500 300
11 # set up topography object
12  set topo [new Topography]
13  $topo load_flatgrid 500 500
14 # Create God
15  create-god 2
16 # Create channel
17  set wChan1 [new Channel/WirelessChannel]
18 # configure a mobile node,
19 $ns node-config -adhocRouting DSDV \
20     -llType  LL \
21     -macType Mac/802_11 \
22     -ifqType Queue/DropTail/PriQueue \
23     -ifqLen 50 \
24     -antType Antenna/OmniAntenna \
25     -propType Propagation/TwoRayGround \
26     -phyType Phy/WirelessPhy \
27     -topoInstance $topo \
28     -agentTrace ON \
29     -routerTrace ON \
30     -macTrace OFF \
31     -movementTrace OFF \
32     -channel $wChan1
33 #create nodes
34  set n0 [$ns node]
35  set n1 [$ns node]
36 #Disable Random Motion
37  $n0 random-motion 0
```

```
38  $n1 random-motion 0
39  #node position
40  $n0 set X_ 0
41  $n0 set Y_ 0
42  $n1 set X_ 400
43  $n1 set Y_ 250
44  #size of nodes in NAM
45  $ns initial_node_pos $n0 30
46  $ns initial_node_pos $n1 30
47  # Setup CBR traffic flow between nodes
48  set udp [new Agent/UDP]
49  set null [new Agent/Null]
50  $ns attach-agent $n0 $udp
51  $ns attach-agent $n1 $null
52  $ns connect $udp $null
53
54  set cbr [new Application/Traffic/CBR]
55  $cbr attach-agent $udp
56  $ns at 1.0 "$cbr start"
57
58  #Configuring Node Movements
59  # n1 starts to move
60  $ns at 2.0 "$n1 setdest 100.0 70.0 35.0"
61  # n0 starts to move
62  $ns at 10.0 "$n0 setdest 50.0 18.0 1.0"
63  # Again n1 starts to move away from n0
64  $ns at 15.0 "$n1 setdest 480.0 280.0 15.0"
65
66  # Tell nodes when the simulation ends
67  $ns at 30.0 "$n0 reset";
68  $ns at 30.0 "$n1 reset";
69  $ns at 30.0 "stop"
70  $ns at 30.01 "puts \"NS EXITING...\" ; $ns halt"
71  proc stop {} {
72      global ns tracefd namtrace
73      $ns flush-trace
74      close $tracefd
75      close $namtrace
76      exec nam wrls.nam &
77      exit 0
78  }
79  puts "Starting Simulation..."
80  $ns run
```

The visualization output of Listing 7.2 execution is depicted in NAM snapshots shown in Figures 7.4 and 7.5. It can be seen from Figure 7.4 that initially nodes are far apart, in which case communication between them is not possible. After some time node1 moves closer to node0 due to line 57 in Listing 7.2 such that they come inside each other's communication range, so data transmission is possible between them, as shown in Figure 7.5. Again, after some more time the node moves away from node 0 due to line 61 of the program. When node 1 moves out of the communication range of node 0, data transmission stops again. Initially, the nodes are far apart, as shown in Figure 7.4 and communication is not possible, as they don't come into each

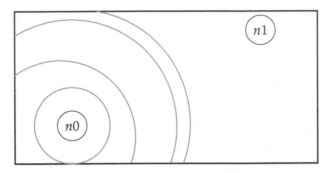

Figure 7.4: Initial position of both nodes; communication is not possible

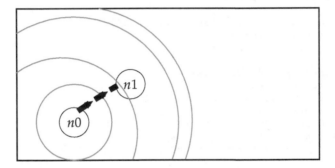

Figure 7.5: Nodes come closer to each other and communication becomes possible

others communication range. But at 2 sec n1 moves toward n0 and and at 10 sec n0 moves toward n1 due to line 57 and line 59 of the above program. Around 12 sec both come within communication range of each other.

Figure 7.5 shows that communication is now possible, as they come sufficiently closer to each other. At 15 sec, again node 1 moves away from node 0 due to line 61. After some time, it goes out of the communication range of node 0 and data communication breaks.

Problem 4: Multihop Communication

If two wireless nodes are not positioned within the communication range of each other, then data transmission is not possible between them. However, if a third node is present such that it is in the communication range of the two nodes, then data communication becomes possible between the two nodes through the third node. That is, the third node will act as a router to forward packets between the two nodes. This can be generalized to make communication possible between two nodes through a large number of intermediate nodes. This technique is called multihop communication. That is, a node can send packets to another node over multiple hops. It is the responsibility of the routing algorithm to determine an appropriate path to the destination. The nodes may behave as ordinary nodes while sending and receiving packets or they may behave as a router when forwarding packets received from a source to their destinations. A program to demonstrate this technique is given in Listing 7.3.

Listing 7.3: Multihop communication

```tcl
1   #---multihop.tcl
2   # Simulates multihop wireless network with 3 nodes
3   #Event scheduler
4   set ns  [new Simulator]
5
6   #trace files
7   $ns use-newtrace
8   set tracefd    [open wrls.tr w]
9   $ns trace-all $tracefd
10  set namtrace [open wrls.nam w]
11  $ns namtrace-all-wireless $namtrace 500 300
12
13  # set up topography object
14  set topo   [new Topography]
15  $topo load_flatgrid 500 300
16
17  # Create God
18  create-god 3
19
20  # Create channel
21  set wChan [new Channel/WirelessChannel]
22
23  # configure mobile node parameters
24  $ns node-config -adhocRouting DSR \
25       -llType   LL \
26       -macType Mac/802_11 \
27       -ifqType CMUPriQueue \
28       -ifqLen 50 \
29       -antType Antenna/OmniAntenna \
30       -propType Propagation/TwoRayGround \
31       -phyType Phy/WirelessPhy \
32       -topoInstance $topo \
33       -agentTrace ON \
34       -routerTrace ON \
35       -macTrace ON \
36       -movementTrace OFF \
37       -channel $wChan
38
39  #create nodes
40   set n0 [$ns node]
41   set n1 [$ns node]
42   set n2 [$ns node]
43  $n0 random-motion 0
44  $n1 random-motion 0
45  $n2 random-motion 0
46  #node color
47  $n0 color blue
```

```
48  $n1 color red
49  $ns at 0.0 "$n0 color blue"
50  $ns at 0.0 "$n1 color red"
51  #node position
52  $n0 set X_ 30
53  $n0 set Y_ 40
54  $n0 set Z_ 0
55  $n1 set X_ 200
56  $n1 set Y_ 100
57  $n1 set Z_ 0
58  $n2 set X_ 350
59  $n2 set Y_ 100
60  $n2 set Z_ 0
61  #size of nodes in NAM
62  $ns initial_node_pos $n0 20
63  $ns initial_node_pos $n1 30
64  $ns initial_node_pos $n2 20
65  # Setup traffic flow between nodes
66  # TCP connections between n0 and n2
67  set tcp [new Agent/TCP]
68  $tcp set class_ 2
69  set sink [new Agent/TCPSink]
70  $ns attach-agent $n0 $tcp
71  $ns attach-agent $n2 $sink
72  $ns connect $tcp $sink
73  set ftp [new Application/FTP]
74  $ftp attach-agent $tcp
75  $ns at 3.0 "$ftp start"
76  # Tell nodes when the simulation ends
77      $ns at 30.0 "$n0 reset";
78      $ns at 30.0 "$n1 reset";
79      $ns at 30.0 "$n2 reset";
80      $ns at 30.0 "stop"
81      $ns at 30.01 "puts \"NS EXITING...\" ; $ns halt"
82
83  proc stop {} {
84      global ns tracefd namtrace
85          $ns flush-trace
86          close $tracefd
87          close $namtrace
88      exec nam wrls.nam &
89      exit 0
90  }
91
92      puts "Starting Simulation..."
93      $ns run
```

The NAM snapshots are given in Figures 7.6 7.8. Figure 7.6 shows the communication range of all three nodes, n0, n1 and n2. As can be seen in Figure 7.6, the communication range of n1 covers both nodes n0 and n1. The communication range of n0 and n2 covers n1. Therefore communication between n0 and n2 is possible via n1. The communication is shown in Figure 7.7, where data packets from n0 are routed by n1 to n2. Figure 7.8 shows the return of ACKs from n2 to n0 through the forwarding node n1.

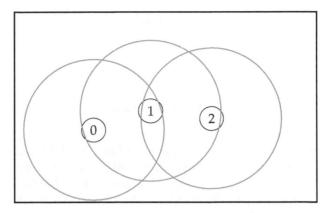

Figure 7.6: Communication range of different nodes shown as circles.

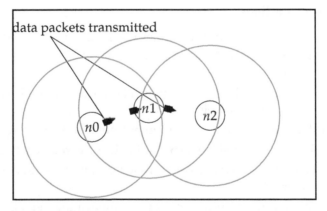

Figure 7.7: Data packets from n0 to n2 are routed through node n1

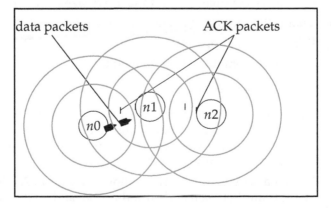

Figure 7.8: ACK packets received from n2 to n0 through n1

Figure 7.9: Three-node wireless network

Communication without making use of any fixed infrastructure is called wireless *ad hoc* networks because this kind of network can be established on the fly as available end nodes in an area form the network and some nodes behave automatically as routers to forward messages. This network is also called infrastructure-less network as opposed to infrastructure-based networks, where routers/base stations are essential parts of the network.

Test/Viva Questions

    a) Why are links between nodes not created in a wireless network?

    b) What is the difference between fixed and mobile wireless networks?

    c) What is the difference between an infrastructure-based and infrastructureless wireless network?

    d) How is a mobile node different from a node in a fixed network?

    e) Why is the CSMA/CD not suitable for wireless networks?

    f) Why can't fixed network routing technology be used in wireless networks?

    g) What is the requirement of two propagation models?

    h) What is the GOD object?

    i) What is the command used to set node size in NAM?

    j) What is the command to create a node movement?

    k) What is a multihop network?

    l) What is the command to color a node?

Programming Assignments

    1) Design a wireless network as shown in Figure 7.9. Now attach a CBR traffic between node 0 and node 1. Attach a movement to node 1. When node 1 moves away from node n0, such traffic should routed through node 2.

    2) Design a wireless network as shown in Figrue 7.9. Now attach a CBR traffic between node 0 and node 2 and an ftp traffic between node 1 and node 3. Position the nodes in such a way that both traffics are forwarded by node 4. Use DSDV protocol.

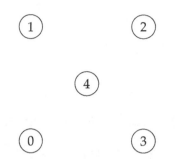

Figure 7.10: Five-node wireless network

3) Design a wireless network as shown in Figure 7.10. Now attach traffic as in assignment 2. Now add movements to the simulation such that node 0 moves to the position of node 1, node 1 moves to node 2's position, node 2 moves to node 3's position, and node 3 moves to node 0's position.

7.3 Wireless Networking Modules

The different modules of NS that facilitate wireless communication are discussed briefly in the following.

7.3.1 Mobile node architecture

A mobile node participating in a packet transmission may behave either as a router or as the destination. A node consists of following network components, as shown in Figure 7.11.

i) A link layer module (LL)

ii) An ARP module connected to the LL (ARP)

iii) An interface priority queue (IFq)

iv) A medium access control layer (MAC)

v) A network interface module (netIF)

vi) All of the above connected to the channel

Link Layer Module (LL)

When a packet is sent from a mobile node (outgoing packet), the packet is transferred down to the LL by the routing agent. The LL then transfers the packet to the interface queue. Similarly, when a packet is received by the mobile node (incoming packet), the MAC layer transfers the packet up to the LL, which in turn transfers it to the node_entry_ point. It is then either processed by the routing agent (if forwarding node) or consumed by the receiver (sink).

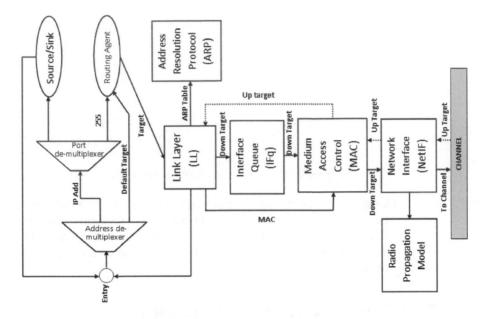

Figure 7.11: Mobile node architecture in NS

Address Resolution Protocol Module (ARP)

When a packet is sent from a mobile node, the LL module puts a query to the ARP module to obtain the MAC address of the next hop. If the ARP already has the MAC address of the next hop, it directly writes it into the MAC header of the packet. Otherwise, it broadcasts an ARP query and caches the packet temporarily. Once the required MAC address is obtained through an ARP reply, the packet is inserted into the interface queue after writing the obtained address into the MAC header of the outgoing packet.

Interface Queue Module (IFQ)

A priority queue (PriQueue) is used as an interface queue. This queue inserts packets at the head of the queue.

Medium Access Control Module (MAC)

IEEE 802.11 and TDMA wireless MAC protocols are normally used in NS.

Network Interface Module (netIF)

This module simulates a hardware interface that is used by a mobile node to access the channel. This interface handles collisions and receives packets transmitted by other nodes through the channel. The interface writes the metadata like the transmission power, wavelength, etc., into each transmitted packet.

The propagation model of the receiving network interface module then uses this metadata available in the received packet header to determine if the packet has enough power to be received by the receiving node. This model implements the *direct-sequence spread-spectrum* (DSSS) radio interface.

Radio Propagation Module (radio)

As discussed earlier, there are two types of radio models. If communicating nodes are at a small distance, then the *Friss-space* model ($1/r2$) is used. The *Two-ray Ground* ($1/r4$) model is used if they are far apart. The model assumes a flat ground plane. Further, mobile nodes use an omnidirectional antenna with unity gain.

7.4   Wireless Routing

Routing protocols designed for a wired network are not suitable in wireless networks because of the medium characteristics. Therefore many routing protocols have been developed for wireless mobile networks over the last two decades. To use any particular routing protocol in a simulation, refer to line 17 of Listing 7.1, where the command is

```
$ns node-config -adhocRouting DSDV
```

To use a specific routing protocol, the DSDV in the above command may be changed with an appropriate routing technique. Several routing techniques for wireless/ad hoc networks have been developed over the last two decades. Many of them are implemented in NS by researchers around the globe and are available as add-ons. A few of them come with the NS installation package and others may be downloaded from various web resource sites.

7.4.1   Destination sequenced distance vector (DSDV)

This is an extension of the distance vector routing scheme used in wired networks. In this technique, routing update messages are exchanged between neighbor nodes. A node is said to be a neighbor of another node if both of them are present in each other's communication range. Two types of routing updates are used, *triggered update* and *routine update*. Updates are triggered if a topology change occurs due to node mobility or link/node failure. Otherwise updates are broadcast in regular time intervals called routine updates. Initially, when the sender wants to send a packet to some destination and the route is not known to it, then a routing query process is initiated, and the packet is cached. Caching of packets continues until a route reply arrives and if the buffer overflows, then packets are dropped using the tail drop mechanism.

Packets destined for a node are sent to the port de-multiplexer and are subsequently handed over to the routing agent, as shown in Figure 7.11. A routing agent is attached in mobile nodes using port number 255. A default-target is used by the mobile node in its classifier (or address demux) and the packet is handed over to this default target when the target is not found for the destination in the classifier (in the case of a forwarding node). The default target is the routing agent and assigns the next hop for the packet and hands the packet down to the LL.

All the routing protocols have been implemented in C++. The code related to the DSDV routing protocol may be found in the *dsdv* subdirectory and in ~ns/tcl/mobility/dsdv.tcl of the NS installation.

7.4.2   Dynamic source routing (DSR)

This protocol uses a separate mobile node called the SR node. The architecture of this node differs slightly from the mobile node as shown in Figure 7.12.

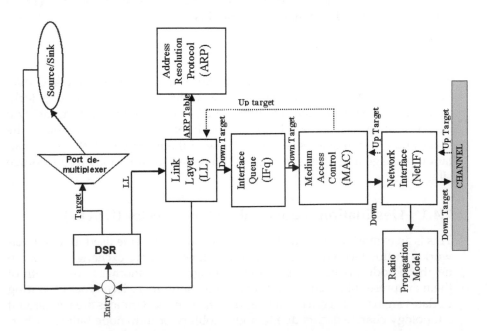

Figure 7.12: Mobile SR node architecture in NS

Unlike other routing protocols, DSR does not use routing table for packet forwarding. Rather, when any node needs to send a packet, it initiates a *route discovery* mechanism in which it tries to find a route toward the destination by flooding a *route request* (RREQ) message. The flooded RREQ packets pass through various nodes, record the route in the packet headers, and eventually reach the destination. Now by analyzing the header information of the received packets, the destination responds with a unicast *route reply* (RREP) message with optimized route information. When the source gets back the RREP, it extracts the route information and writes the information into each outgoing data packet.

Now to forward a packet toward the destination, the routing agent checks the packet header where the *source route* (routing path) information is available. If the source route information is not found in the packet header, then the agent checks if the information is available with itself; if it is, then it provides the source route to the packet. Otherwise it caches the packet and initiates route discovery by broadcasting RREQ packets.

It can be observed that the main working principle of DSR is as follows.

Routing queries are initiated when a data packet contains no route to its destination. Initially, these RREQs are broadcast to all their neighbors. These neighbors further broadcast these RREQs if they do not have the route information. In this process, route replies may come either from intermediate nodes or from the destination node, to the source. Then the routing agent hands over all the packets that are destined to itself to the port demultiplexer, as shown in Figure 7.12.

The implementation of the DSR algorithm is available in the *dsr* directory and in ∼ns/tcl/mobility/dsr.tcl.

7.4.3   Temporally ordered routing algorithm (TORA)

TORA is a distributed routing protocol designed for wireless networks and is based on a *link reversal* algorithm. In each node, different instances of the TORA protocol run for different destinations. When any node requires a route to some arbitrary destination, it then floods a QUERY message containing the destination address for which it needs a route. This query packet travels through the network until it reaches the destination (if connected) or until it reaches an intermediate node that already has a route to the required destination node.

When a node having information about the destination receives a QUERY packet, it broadcasts an UPDATE packet. The UPDATE packet contains its height to the destination. Height is the path length or number of hops required to reach the destination. As this packet travels toward the source, each intermediate node receiving this message updates its own height. This results in a set of paths (directed links) from the source to the destination that may be of different heights.

If a node finds a destination to be unreachable, then it sets a local maximum value of height for that particular destination. If a node fails to find any neighbor having a finite height to a destination, then it initiates another query process to find a new route. When a network partition is detected, the detecting node broadcasts a CLEAR message which resets all routing states.

The implementation of TORA is found in the *tora* directory, usually located at ns/tcl/mobility/tora.tcl.

7.4.4   Ad hoc on demand distance vector (AODV)

AODV combines ideas from both DSR and DSDV protocols. It applies the *routediscovery* and *routemaintenance* techniques of DSR and uses the hop-by-hop routing, sequence numbering, and beacons from the DSDV protocol. When a node needs a route to any destination, it generates an RREQ, This forwarded by intermediate nodes and helps create a reverse route to the source. If an intermediate node receiving the request already has the route to the destination, it generates a route reply (RREP) packet containing the number of hops required to reach the destination from the current node and sends it to the source node. All the nodes participating in forwarding the reply to the source node also store a forward route to the destination.

The implementation of AODV is found in the *aodv* directory, usually located at ~ns/tcl/lib/ns-lib.tcl.

7.5   Wireless Trace

NS provides two kinds of outputs for any simulation. One is the NAM trace, which is used by the NAM package to provide animation of the simulation, another is the event trace, which records each event of the simulation row wise in a file. Normally, the NAM trace is stored in a file with extension *<fileName>.nam* and the event trace file's extension is *<fileName>.tr*. Up to this point we have discussed the NAM trace only. The event trace is more important, as it is useful to measure the performance of the network. To do so we need to understand the format of the event trace. There is a old format and a new format available for wireless networks. We will discuss the new version of the trace format. The new version of the trace may be generated using the following command.

```
$ns use-newtrace
```
And this command needs to be used before the trace command
```
$ns trace-all <trace-fd>
```
When the simulation finishes, the trace file produced stores all the events generated in the simulation as one event in one row. A few events as stored in a wireless simulation event trace file are given as follows.

Event 1
```
s -t 65.1 -Hs 1 -Hd -1 -Ni 1 -Nx 122.46 -Ny 117.46 -Nz 0 -Ne
-1.0
-Nl RTR -Nw -- -Ma 0 -Md 0 -Ms 0 -Mt 0 -Is 1.255 -Id -1.255
-It message -Il 32 -If 0 -Ii 15 -Iv 32
```
Event 2
```
r -t 65.13 -Hs 0 -Hd -1 -Ni 0 -Nx 20.00 -Ny 18.00 -Nz 0.0 -Ne
-1.0
-Nl MAC -Nw -- -Ma 0 -Md ffffffff -Ms 1 -Mt 800 -Is 1.255 -Id
-1.255
-It message -Il 32 -If 0 -Ii 15 -Iv 32
```
Event 3
```
d -t 65.135295138 -Hs 0 -Hd 1 -Ni 0 -Nx 20.00 -Ny 18.00 -Nz
0.00
-Ne -1.000000 -Nl IFQ -Nw ARP -Ma 0 -Md 0 -Ms 0 -Mt 800 -Is
0.0
-Id 1.0 -It tcp -Il 80 -If 2 -Ii 2 -Iv 32 -Pn tcp -Ps 0 -Pa 0
-Pf 0 -Po 0
...
```

In one trace file there will be hundreds/thousands of these lines written, but with a common format. The various types information are stored as a $-Tag$ $value$ pair in different columns/fields. Each of these $-Tag$ $value$ pairs taken from the last event of the above example are shown below. All 27 pieces of information provided below belong to one event only, and all the events contain similar information. So, if we understand one event, then we can extract the required information from multiple events available in a trace file using AWK scripting, as we did in the last chapter.

1. $d^{(1)}$ $\rightarrow$ Event Type
2. $-t^{(2)}$ $65.135295138^{(3)}$ $\rightarrow$ Time of Event

Next-Hop Information

3. $-Hs^{(4)}$ $0^{(5)}$ $\rightarrow$ ID of this Node
4. $-Hd^{(6)}$ $1^{(7)}$ $\rightarrow$ ID of next hop towards Destination

Node Property Tags

5. $-Ni^{(8)}$ $0^{(9)}$ $\rightarrow$ Node id
6. $-Nx^{(10)}$ $20.00^{(11)}$ $\rightarrow$ Node's x-coordinate
7. $-Ny^{(12)}$ $18.00^{(13)}$ $\rightarrow$ Node's y-coordinate
8. $-Nz^{(14)}$ $0.00^{(15)}$ $\rightarrow$ Node's z-coordinate
9. $-Ne^{(16)}$ $-1.000000^{(17)}$ $\rightarrow$ Node energy level
10. $-Nl^{(18)}$ $IFQ^{(19)}$ $\rightarrow$ Trace level
11. $-Nw^{(20)}$ $ARP^{(21)}$ $\rightarrow$ Reason for the drop event

MAC level Packet Information

12. $-Ma^{(22)}$ $0^{(23)}$ $\rightarrow$ Duration
13. $-Md^{(24)}$ $0^{(25)}$ $\rightarrow$ Destination's Ethernet

```
         address
14.     −Ms(26)  0(27) → Source's Ethernet address
15.     −Mt(28)  800(29) → Ethernet Type
```

IP level Packet Information

```
16.     −Is(30)  0.0(31) → Source Address.Source
        Port Number
17.     −Id(32)  1.0(33) → Destination Address.Desti-
        nation Port Number
18.     −It(34)  tcp(35) → Packet Type
19.     −Il(36)  80(37) → Packet Size
20.     −If(38)  2(39) → Flow id
21.     −Ii(40)  2(41) → Packet unique id
22.     −Iv(42)  32(43) → TTL value
```

Application level Packet Information

```
23.     −Pn(44)  tcp(45) → Info about TCP flows
24.     −Ps(46)  0(47) → Packet sequence number
25.     −Pa(48)  0(49) → ACK number
26.     −Pf(50)  0(51) → Number of times the packet
        is forwarded
27.     −Po(52)  0(53) → Optimal number of forwards
```

One event trace is provided in tabular format above, where the first column represents the Tag and the second column provides the value for that Tag. The numbers in parenthesis represent the column number in the original event trace.

Column 1 represents event type. The various event types possible are

```
s: send
r: receive
d: drop
f: forward
```

Columns 2 and 3 provide time of the event. Columns 4 to 7 provide next-hop information. Columns 8 to 21 are called node property tags. Column 11 gives the reason for a packet drop. The various reasons possible for a packet drop are as follows.

```
END: drop due to end of simulation
COL: drop due to collision at MAC layer
DUP: drop due to duplicate packet
ERR: drop due to MAC packet error
RET: drop due to exceeding retry count
STA: drop due to MAC invalid state
BSY: drop due to MAC busy
NRTE: drop due to no route available
LOOP: drop due to a routing loop
TTL: drop due to TTL = 0
TOUT: drop due to packet expired
```

```
IFQ: drop due to no buffer space in IFQ
ARP: dropped ARP
OUT: dropped by base stations
```

Columns 22 to 29 represent packet information at the MAC level. Columns 30 to 42 provide packet information at the IP level. Columns 44 to 52 provide packet information at the Application level. Different Applications have different traces as follows.

-Pn arp: ARP information
        -Po: ARP request/reply
        -Pm: Source MAC address
        -Ps: Source IP address
        -Pa: Destination MAC address
        -Pd: Destination IP address
-P dsr: DSR information
        -Pn: number of nodes traversed
        -Pq: route request flag
        -Pi: route request sequence number
        -Pp: route reply flag
        -Pl: reply length
        -Pe: source of routing -> destination
        of routing
        -Pw: error flag
        -Pm: number of error
        -Pc: whom to report?
        -Pb: link error from link a-> link b
-Pn cbr: CBR information
        -Pi: packet sequence number
        -Pf: number of times packet is forwarded
        -Po: optimal number of forwards
-Pn tcp: TCP information
        -Ps: packet sequence number
        -Pa: ACK number
        -Pf: number of times packet is forwarded
        -Po: optimal number of forwards

All the information provided above is useful to extract the required information from an event trace file. The complete methods are discussed in the next section.

7.6 Network Performance Metrics

There can be many metrics designed to measure the performance of a network. Some of the major ones are throughput/bandwidth, delay, packet loss, jitter, etc. All these values for a particular simulation can be found from the event trace file generated from an NS simulation. The methods for some of the metrics are given in this section. Let us rewrite the Listing 7.2 to generate a valid trace file. Add the line

```
$ns use-newtrace
```

before line 3 of Listing 7.2, so that the trace generated will be in a new format. When the simulation is executed, it generates a trace file named *wrls.tr*. You may open the file using any text editor to see the content. It contains several hundred thousand lines based on simulation type and simulation time. To measure different network performances we need to write different AWK scripts to be run over the trace file generated, as given in the rest of this section.

i. **Network Throughput**

This is defined as the total number of bytes received by the destination per sec. To calculate this, we use following two simple steps.
1. Add the sizes of all data packets received.
2. Divide the total size calculated in the above step by the simulation time.
The AWK script for the above procedure is given in Listing 7.4.

Listing 7.4: Wireless throughput

```awk
1  #======= wLessThPut.awk =======
2  #!/bin/awk -f
3  # Script to find out throughput in a wireless network.
4  # Run command: awk -f wLessThput.awk < wrls.tr
5
6  BEGIN  {
7      totBytesRecvd = 0
8      simTime=30
9  }
10 {
11   # initialize required variables
12   event = $1
13   pktSize = $37
14       pktType = $19
15       # Calculate all received data packets' size in bytes
16   if  (event == "r" && pktType == "AGT") {
17     #print $1 $2 $3
18           totBytesRecvd += pktSize
19       }
20 }
21 END {
22       printf("\n Network Throughput = %g kbps \n", (
         totBytesRecvd / simTime)*(8/1000))
23 }
```

Output:
Network Throughput = 189.581 kbps

ii. **Packet Loss**
Packet loss is defined as the number of packets not received out of

total packets sent. More specifically, we are interested in packet loss percentage, which counts the number of packets lost out of each 100 packets sent. The formula may be written as

$$Packet\ loss = \frac{Packets\ dropped}{Packets\ transmitted} * 100\ \%$$

The AWK script to find the packet loss percentage is given in Listing 7.5.

Listing 7.5: Percentage packet loss

```awk
1  #======= pktLoss.awk ======
2  #!/bin/awk -f
3  # Script to find out packets loss percentage.
4  # Run Command: awk -f pktLoss.awk < wrls.tr
5  BEGIN  {
6    for  (i in pktsSent) {
7           pktsSent[i] = 0
8    }
9    for (i in recv) {
10          pktsRecvd[i] = 0
11   }
12        txed = 0
13        drop = 0
14        pktLoss = 0
15 }
16 {
17   event = $1
18       pktSeq = $47
19   pktType = $19
20
21       # If a packet is transmitted then set it
22       if ( pktsSent[pktSeq] == 0 &&  event == "s"  &&
     pktType == "AGT") {
23           pktsSent[pktSeq] = 1
24       }
25       # If a packet is received then set it
26       if (event == "r" && pktType == "AGT") {
27           pktsRecvd[pktSeq] = 1
28       }
29 }
30 END {
31   # find total number of packets transmitted
32       for (i in pktsSent) {
33           if  (pktsSent[i] == 1) {
34                txed ++
35           }
36           # find total number of packets received/ dropped
37       if (pktsRecvd[i] == 0 ) {
38                drop ++
39       }
40       }
41       # find the ratio of dropped packet to transmitted
     packet
42       if (txed != 0) {
```

```
43            pktLoss = drop / txed
44        } else {
45            pktLoss = 0
46        }
47        printf("Pkts Txeded = %g, Pkts Dropped = %g\n Pkt
        Loss = %g %%\n", txed,drop,pktLoss*100)
48 }
```

Output:
Pkts Txed = 7734, Pkts Dropped = 4644
Pkts loss = 60.04 %

iii. **Average Delay**
   This is the difference between the receiving and sending times of a
   packet, averaged for each packet received. The average delay is calculated using the following steps.
   1. For each packet i received, find the delay as
   $delay_i = arrivalTime\_Packet_i - sentTime\_Packet_i$.
   2. $avgdelay = \frac{\sum_{i=1}^{n} delay_i}{n}$
   The AWK code to find the average delay is given in Listing 7.6.

Listing 7.6: Average delay

```
1  # ====== avgDelay.awk ======
2  #!/bin/awk -f
3  # Script to find out average Delay occurred in a wireless
       network.
4  # Run Command: awk -f avgDelay.awk < wrls.tr
5  BEGIN  {
6      for (i in pktSent) {
7          pktSent[i] = 0
8      }
9      for (i in pktRecvd) {
10         pktRecvd[i] = 0
11     }
12     delay = avgDelay = 0
13 }
14 {
15     event = $1
16     time = $3
17     pktSeq = $47
18     pktType = $19
19     # Store packets sent time
20     if (pktSent[pktSeq] == 0 &&  event == "s" && pktType ==
       "AGT") {
21        pktSent[pktSeq] = time
22     }
23     # Store packets arrival time
24     if (event == "r" && pktType == "AGT") {
25        pktRecvd[pktSeq] = time
26     }
27 }
28 END {
```

```
29      # Compute average delay
30   for (i in pktRecvd) {
31        if (pktSent[i] == 0) {
32            printf("\n Error %g\n",i)
33        }
34        delay += pktRecvd[i] - pktSent[i]
35        num ++
36   }
37   if (num != 0) {
38       avgDelay = delay / num
39   } else {
40       avgDelay = 0
41   }
42   printf("Avg Delay = %g ms\n",avgDelay*1000)
43 }
```

Output:
Avg Delay = 3.54164 ms

Test/Viva Questions

a) What is the use of the ARP module in a mobile node architecture and it is attached to which module?

b) For an outgoing packet, the routing agent forwards the packet to which module?

c) What are the different protocols available for the MAC layer (module)?

d) The propagation module is connected to which module?

e) What are the differences between free-space and two-ray ground propagation? How do we choose one?

f) What are the architectural differences between a common mobile node and an SR node?

g) What is the command to generate traces according to the new trace format?

h) Which tags of a trace file represent the node's coordinates?

i) What are the various event types and which tag is used to represent them in a trace file?

j) Which tag is responsible for packet size?

k) Which tag is used find the trace level?

l) Which tag is used for packet sequence number in TCP?

m) Which tag is used to find the next hop ID?

n) What is the throughput of a network? How do we calculate it?

o) How do we find a packet loss from a trace file?

Programming Assignments

1) Rewrite Listing 7.2 with a TCP agent and ftp traffic. Measure the performances like throughput, packet loss, and average delay. Now compare these values with the values measured for CBR traffic by plotting bar graphs.

2) Execute the program in Listing 7.2 three times, each time using a different routing protocol (DSR, DSDV, AODV). Measure the performance metrics and plot suitable graphs.

3) Simulate a wireless network as shown in Figure 7.9. Attach an ftp traffic between n0 and n1 that starts at 0.1 sec. Now n1 moves away from n0 at 3 sec such that data communication between n0 and n1 is performed via n2. Run the simulation for 50 sec and calculate throughput, delay, and packet loss.

4) Simulate a wireless network as shown in Figure 7.10. Attach two ftp traffics between n0 and n2 and n1 and n3 respectively. Both traffics are forwarded through n4. Run the simulation for 50 sec and calculate

    i. flow-wise throughput, delay, and packet loss.

    ii. overall network throughput, delay, and packet loss.

    Hint: use the *set fid* command to distinguish the flows in an event trace file. Columns 38 and 39 of the trace file are used to represent flow ID.

5) Design a wireless simulation with six nodes in an area 500 m × 700 m. Attach two FTP traffics and two CBR traffics to any arbitrary source and destination. Apply random motion to all nodes with DSR routing protocols. Run the simulation for 150 sec. Measure throughput and delay metrics for different flows separately. Plot four different throughputs in a single bar chart and four different delays in another bar chart with proper labels.

7.7 Practical Simulation Issues

Up to this point, we have seen simulations involving two to five wireless nodes, but in real-life wireless network simulation, it may be required to simulate networks containing nodes on the order of 10s or 100s or even 1000s. In that case, creating the nodes manually, defining their initial individual initial positions and mobility patterns, creating connections between different nodes, and attaching transport and traffic to each connection become tedious. In NS, there is a facility to perform these tasks automatically by using some predefined scripts, which perform all of the above tasks by taking random values in the requested range of vales. The following sections describe the sequence of steps to perform these activities.

7.7.1 Generating scenario

There is a script named *setdest* available under the directory ∼*ns/indep-utils/cmu-scen-gen/setdest* that creates initial topology by defining the initial position of the nodes in the given area and node movement to random points in random times.

Run the *setdest* command with the following arguments.

```
<path>/setdest -v <2> -n <nodes> -s <speed type> -m
<min speed> -M <max speed> -t <simulation time> -P
<pause type> -p <pause time> -x <max X> -y <max Y> >
<outdir>/<scen-file>
```

The values for the arguments (given in angular brackets) need to be supplied by the user. The meanings of the arguments above are as follows.

-v : version number (1 or 2)

-n : number of wireless nodes in the simulation

-s : speed type

-m : minimum speed of the nodes in m/s

-M : maximum speed of the nodes in m/s

-t : total time of simulation

-P : type of pause

-p : pause time

Nodes in each scenario move according to the random waypoint model in which each node individually selects a destination coordinate (point) randomly in the flat test bed. It also selects the speed of movement from a given interval (0, maxSpeed) and moves toward the destination point at the selected speed. After reaching the destination point, it stops for *pause time* seconds and then repeats the process. The *pause time* 0 is considered continuous movement of nodes. -x, -y define the maximum value for the x and y coordinates. In other words, these values represent the coordinate values for the upper right corner point of the considered rectangular area of simulation.

Besides these arguments, <path> defines the directory in which the program *setdest* is available. Finally, at the end <outdir>/<scen-file> defines the output file, where the output of the above command is stored and may be used in the simulation. We will see the actual use of this command in the example given below.

7.7.2 Generating a connection pattern

Once the scenario is generated using the above command, we need to define one or more connections between the nodes. A connection is nothing but a flow between an arbitrary set of source and destination nodes. It involves, choosing the source and destination nodes, attaching transport agents and application traffic to them, and finally defining start and stop times for the traffic. It requires writing 8 to 12 lines of TCL code for each connection. If we are required to define connections on the order 10s or 100s, then a huge amount of code must be written by the programmer, which is redundant. To avoid this situation, the TCL script available under the directory ~ns/tcl/mobility/scene/cbr may be used as given below.

```
ns <path>/cbrgen.tcl [type cbr|TCP] [-nn nodes] [-seed seed]
        [-mc connections] [-rate rate] >
<outdir>/<conn-file>
```

Various arguments passed to the scripts are defined as follows.
type: CBR traffic or TCP traffic
-nn: total number of nodes available in the simulation
-seed: seed to generate random times required to define start and stop times for the traffic flow
-mc: number of connections (flows) to be created
-rate: rate of packet flow (number of packets transmitted per sec)

The output of the above script is stored in *conn-file*, which may be used in a simulation.

Before using the above two commands, some important points to be considered are the following.

i) The number of connections supplied to the command must be less than or equal to half the number of nodes present in the simulation. This is because, the command is designed such that it considers different nodes for a new connection. That is, if two nodes are already used for a connection, then neither of these nodes can be used for the next connection. Therefore, if you need two connections, there must be at least four nodes present in the simulation. In practice, for very large networks simulation, the number of connections is kept much lower than half the nodes.

ii) The value for the *rate* argument above must be a floating point number (4.0 is OK, not 4). Since the packet interval is calculated as 1/rate, therefore, if an integer value is given for rate, then the packet interval becomes 0.

iii) If the type for above command is CBR, then it attaches a UDP agent to nodes and CBR traffic is assigned. If TCP is chosen, it assigns FTP traffic on the top of the TCP agent.

However, these points may not be considered as limitations of the above command. If you require different behavior, such as the same node containing more than one flow, or the rate to be defined as integers, etc., this can always be achieved by modifying the code for the above commands.

Let us use these commands for a real network simulation. We keep the example small (three nodes and one connection) in order to make it simple to understand and to keep it within the limited scope of the text, but it can always be scaled to any required size. The command to generate a scenario for three nodes within a rectangular simulation area is given as follows.

```
~/ns-allinone-2.35/ns-2.35/indep-utils/
cmu-scen-gen/setdest/setdest -v 2 -n 3 -s 1 -m 1 -M 19 -t 100
-P 1 -p 50 -x 300 -y 1500 > scen_3node
```

The above command after execution generates a three-node mobile wireless network within a simulation area of 300 m × 1500 m, nodes move with any speed between 1 m/s and 19 m/s. The pause time is 50 s. The output of the command is redirected to a file named *scen_3node*, which contains the script for initial node position and movement of nodes, etc. The content of the file *scen_3node* is as follows.

```
 1  #
 2  # nodes: 3, speed type: 1, min speed: 1.00, max speed: 19.00
 3  # avg speed: 3.86, pause type: 1, pause: 50.00, max x: 300.00,
       max y: 1500.00
 4  #
 5  $node_(0) set X_ 148.637650165854
 6  $node_(0) set Y_ 843.047099153930
 7  $node_(0) set Z_ 0.000000000000
 8  $node_(1) set X_ 38.506294532984
 9  $node_(1) set Y_ 695.257751413996
10  $node_(1) set Z_ 0.000000000000
11  $node_(2) set X_ 174.248299641452
12  $node_(2) set Y_ 650.726946851054
13  $node_(2) set Z_ 0.000000000000
14  $ns_ at 0.000000000000 "$node_(2) setdest 201.792157155641
       978.216353214579 6.918896395045"
15  $god_ set-dist 0 1 1
16  $god_ set-dist 0 2 1
17  $god_ set-dist 1 2 1
18  $ns_ at 34.802579279568 "$god_ set-dist 1 2 2"
19  $ns_ at 47.499723893694 "$node_(2) setdest 201.792157155641
       978.216353214579 0.000000000000"
20  $ns_ at 50.000000000000 "$node_(0) setdest 195.708136308774
       1226.315206166721 4.200136178552"
21  $ns_ at 50.000000000000 "$node_(1) setdest 84.007254143597
       1117.478350744466 18.224273473641"
22  $ns_ at 54.748143320075 "$god_ set-dist 1 2 1"
23  $ns_ at 73.302177033081 "$node_(1) setdest 84.007254143597
       1117.478350744465 0.000000000000"
24  $ns_ at 97.499723893694 "$node_(2) setdest 261.719521562955
       77.994109765341 14.892619850967"
25  #
26  # Destination Unreachables: 0
27  #
28  # Route Changes: 2
29  #
30  # Link Changes: 2
31  #
```

```
32 # Node / Route Changes / Link Changes
33 #    0 /            0 /               0
34 #    1 /            2 /               2
35 #    2 /            2 /               2
36 #
```

The same command may be used to create a network with any number of nodes by changing appropriate parameter values. Then the connection (flow) between the nodes is generated using the following command.

```
ns ~/ns-allinone-2.35/ns-2.35/indep-utils/
cmu-scen-gen/cbrgen.tcl -type cbr -nn 3 -seed 234567 -mc 1
-rate 4.0 > traffic_3node
```

When executed, the command creates one connection between any two nodes chosen randomly. Remember that the maximum number of connections must be half the number of nodes; therefore we chose to have one connection, as there are only three nodes available. The output of the command is redirected to the file *trafic_3node* and the content is as follows.

```
#
# nodes: 3, max conn: 1, send rate: 0.25, seed: 234567
#
# 1 connecting to 2 at time 8.0821075607473531
#
set udp_(0) [new Agent/UDP]
$ns_ attach-agent $node_(1) $udp_(0)
set null_(0) [new Agent/Null]
$ns_ attach-agent $node_(2) $null_(0)
set cbr_(0) [new Application/Traffic/CBR]
$cbr_(0) set packetSize_ 512
$cbr_(0) set interval_ 0.25
$cbr_(0) set random_ 1
$cbr_(0) set maxpkts_ 10000
$cbr_(0) attach-agent $udp_(0)
$ns_ connect $udp_(0) $null_(0)
$ns_ at 8.0821075607473531 "$cbr_(0) start"
#
#Total sources/connections: 1/1
#
```

These two files are to be used in a simulation script. Again, it is also possible to modify the content of these two files manually before using them in a simulation. For example, the *cbr* parameters given above may be changed to any other valid values, e.g., packet size may be changed from 512 to 1024 etc. The NS program that uses these files is given in Listing 7.7.

Listing 7.7: Simulation using scenario and connection pattern files

```
1 # ---- automated.tcl ------
2 # Program uses connection pattern and scenario files generated
      by
```

```
3   # cbrgen.tcl and setdest programs
4   # ================
5   # Define options
6   # ================
7   set opt(chan)  Channel/WirelessChannel
8   set opt(prop)  Propagation/TwoRayGround
9   set opt(netif)  Phy/WirelessPhy
10  set opt(mac)   Mac/802_11
11  set opt(ifq)   Queue/DropTail/PriQueue
12  set opt(ll)    LL
13  set opt(ant)   Antenna/OmniAntenna
14  set opt(x)     300  ;# X dim of the topography
15  set opt(y)     1500 ;# Y dim of the topography
16  set opt(ifqlen) 50  ;# max packet in ifq
17  set opt(seed)  0.0
18  set opt(tr)    wls_3node.tr   ;# trace file
19  set opt(nam)   wls_3node.nam  ;# nam trace file
20  set opt(rp)    DSDV
21  set opt(nn)    3    ;# number of nodes in simulation
22  set opt(cp) "./traffic_3node" ;# traffic pattern
23  set opt(sc) "./scen_3node"  ;# scenario pattern
24  set opt(stop) 100   ;# simulation time
25
26  # ==============
27  # Main Program
28  # ==============
29  # create simulator instance
30   set ns_ [new Simulator]
31
32  # set topography objects
33   set wtopo [new Topography]
34
35  # create trace object for ns and nam
36   set tracefd [open $opt(tr) w]
37   set namtrace    [open $opt(nam) w]
38   $ns_ trace-all $tracefd
39   $ns_ namtrace-all-wireless $namtrace 300 1500
40
41  # use new trace file format
42   $ns_ use-newtrace
43
44  # define topology
45   $wtopo load_flatgrid $opt(x) $opt(y)
46  # Create God
47   set god_ [create-god $opt(nn)]
48
49  #configure mobile node
50  set chanl [new $opt(chan)]
51  $ns_ node-config -adhocRouting $opt(rp) \
52                   -llType $opt(ll) \
53                   -macType $opt(mac) \
54                   -ifqType $opt(ifq) \
55                   -ifqLen $opt(ifqlen) \
56                   -antType $opt(ant) \
57                   -propType $opt(prop) \
58                   -phyType $opt(netif) \
```

```
59                      -channel $chanl \
60                      -topoInstance $wtopo \
61                      -agentTrace ON \
62                      -routerTrace OFF \
63                      -macTrace ON
64
65   #  Create the specified number of nodes [$opt(nn)]
66   for  {set i 0} {$i < $opt(nn) } {incr i} {
67     set node_($i) [$ns_ node]
68       $node_($i) random-motion 0 ;# disable rand motion
69   }
70
71   # Define node movement model
72   puts "Loading scenario file..."
73   source $opt(sc)
74
75   # Define traffic model
76   puts "Loading connection pattern..."
77   source $opt(cp)
78
79   # Define node size in nam
80   for {set i 0} {$i < $opt(nn)} {incr i} {
81       $ns_ initial_node_pos $node_($i) 40
82   }
83
84   # Tell nodes when the simulation ends
85   for {set i 0} {$i < $opt(nn) } {incr i} {
86           $ns_ at $opt(stop).000000001 "$node_($i) reset";
87   }
88
89   # tell nam the simulation stop time
90   $ns_ at  $opt(stop) "$ns_ nam-end-wireless $opt(stop)"
91   $ns_ at  $opt(stop).000000001 "puts \"NS EXITING...\" ; $ns_
         halt"
92   puts "Starting Simulation..."
93
94   $ns_ run
```

In lines 69 to 74 in the program given in Listing 7.7, we use the movement
and traffic files generated earlier. Execute the program now and observe the
simulation. The simulation is as usual, only we have used an automated scenario and connection generation without writing them manually. Another
modification is that we store various values in an associative array before
assigning them to various parameters in lines 4 to 21. The advantage is that,
if we want to change the value of a parameter and the value is used multiple times in a program, then we need to change it once in the initialization,
otherwise we would have to change this value in each use throughout the
program.

7.7.3   More performance metrics

Along with throughput and delay metrics discussed earlier, the different protocol performance may also be evaluated using the following metrics, which
are computed over the whole duration of the simulation.

i) **Packet Delivery Ratio**: The ratio of data packets sent by the source that are received by the receivers. For example, if there are 2 sources and 2 receivers, and each source sends 10 packets, and one receiver receives 8 packets and another receives 6, the packet delivery ratio would be $(8+6)/(10\times2)$, which is 0.7, i.e., packets actually received/packets supposed to be received.

ii) **Normalized Overhead**: This is the ratio of the number of control and data packet (or bytes) transmitted to the successfully delivered data packet (or bytes). This metric evaluates the overall effort required by the protocol to deliver each data packet. For example, an efficiency of 5 means that the protocol makes 5 packet transmissions on an average for each data packet that is delivered to a receiver.

iii) **Forwarding Efficiency**: This is calculated as the average number of times each originated data packet was forwarded by the intermediate nodes to reach to the destination. This metric represents the efficiency of the routing protocol.

iv) **Delivery Latency**: This is the time difference between the receiving and sending times of a data packet, averaged over all received packets.

7.8 A Complete Example

In most real-life cases, we need to compare the performance of different protocols in different environments. Let us consider a scenario where it is required to evaluate the packet delivery ratio of both DSDV and AODV routing protocols in two situations. One is with different mobility speed and another is with a different network load. Let us define different network simulation parameters as follows.
Common attributes:
            Area: 500 m × 500 m
            nodes: 100
            pause time: 10 sec
            simulation time: 900 sec
            Routing protocols: DSDV and AODV
            Packet rate: 6 pkt/sec
            Traffic: CBR
Scenario 1: Varying mobility speed
            Speed of movement: 1 m/sec to 35 m/sec
            Network load (medium): 25 connections
Scenario 2: Varying network load
            Network load: 5–50 connections
            Speed of movement: 17 m/sec

We solve this problem using the following steps.

Step 1: Writing the simulation script as given in Listing 7.8.

Step 2: Writing AWK scripts to find the packet delivery ratio in both the situations as given in Listings 7.9 and 7.10.

Step 3: Write a shell script that generates a scenario and connection file for both scenarios, and executes the simulation script (Listing 7.8) designed in step 1, and collects the results using AWK scripts designed in step 2. The complete shell script is given in Listing 7.11.

Listing 7.8: Simulation NS script with command line args

```
1  # ---- protoComp.tcl ----
2  # Program is executed automatically by the shell script:
3  # protoconp.sh
4  # ================
5  # # Define options
6  # ================
7  set opt(chan)    Channel/WirelessChannel
8  set opt(prop)    Propagation/TwoRayGround
9  set opt(netif)   Phy/WirelessPhy
10 set opt(mac)     Mac/802_11
11 set opt(ifq)     Queue/DropTail/PriQueue
12 set opt(ll)      LL
13 set opt(ant)     Antenna/OmniAntenna
14 set opt(x)       ""     ;# X dimension
15 set opt(y)       ""     ;# Y dimension
16 set opt(ifqlen)  50     ;# max packet in ifq
17 set opt(tr)      ""     ;# trace file
18 set opt(rp)      ""
19 set opt(nn)      ""     ;# number of nodes
20 set opt(cp)      ""
21 set opt(sc)      ""
22 set opt(time)    ""     ;# simulation time
23
24 # =====================================
25 # # Main Program
26 # =====================================
27 proc getopt {argc argv} {
28   global opt
29     for {set i 0} {$i < $argc} {incr i} {
30       set arg [lindex $argv $i]
31       if {[string range $arg 0 0] != "-"} continue
32         set name [string range $arg 1 end]
33         set opt($name) [lindex $argv [expr $i+1]]
34     }
35 }
36 getopt $argc $argv
37
38 # create simulator instance
39 set ns_ [new Simulator]
40
41 # create trace object for ns and nam
42  set tf [open $opt(tr) w]
43 $ns_ trace-all $tf
44 $ns_ use-newtrace
```

```
45
46  # set up topography object
47   set topo        [new Topography]
48   $topo load_flatgrid $opt(x) $opt(y)
49
50   # Create God
51  set god_ [create-god $opt(nn)]
52
53  # Create channel
54  set wChan [new $opt(chan)]
55
56   $ns_ node-config -adhocRouting $opt(rp) \
57                    -llType $opt(ll) \
58                    -macType $opt(mac) \
59       -ifqType $opt(ifq) \
60                    -ifqLen $opt(ifqlen) \
61                    -antType $opt(ant) \
62                    -propType $opt(prop) \
63                    -phyType $opt(netif) \
64                    -topoInstance $topo \
65                    -agentTrace ON \
66                    -routerTrace OFF \
67                    -macTrace OFF \
68                    -movementTrace OFF \
69                    -channel $wChan
70
71      for {set i 0} {$i < $opt(nn)} {incr i} {
72         set node_($i) [$ns_ node]
73            $node_($i) random-motion 0
74      }
75      # load node movement model
76      source $opt(sc)
77
78      # load traffic model
79      source $opt(cp)
80
81      # prepare the stop procedure
82  proc stop {} {
83    global ns_ tf  opt
84          $ns_ flush-trace
85          close $tf
86  }
87
88  # tell ns the simulation stop time
89   $ns_ at $opt(time) "stop"
90   $ns_ at $opt(time).000000002 "puts \"Simulation done.\" ; $ns_
        halt"
91
92  #st art simulation
93   puts "Starting Simulation..."
94   $ns_ run
```

Listing 7.9: AWK script to find Packet delivery ratio with varying network load

```
1  #====== pdrLoad.awk ======
```

```
2  # used automatically by protocomp.sh
3  BEGIN   {
4      totPktsRecvd = 0
5      totPktsSent = 0
6  }
7  {
8      event = $1
9      pktType = $19
10     if (event == "s" && pktType == "AGT") {
11         totPktsSent ++
12     }
13     if (event == "r" && pktType == "AGT") {
14         totPktsRecvd ++
15     }
16 }
17 END {
18     printf("\t%g",(totPktsRecvd / totPktsSent))   >> "PDRvarLoad.
       dat"
19 }
```

Listing 7.10: AWK script to find Packet delivery ratio with varying mobility speed

```
1  #========== pdrSpeed.awk ============
2  # used automatically by protocomp.sh
3   BEGIN {
4    totPktsRecvd = 0
5    totPktsSent = 0
6   }
7   {
8    event = $1
9    time = $3
10   flowID = $39
11   pktSize = $37
12   pktType = $19
13   if (event == "s" && pktType == "AGT") {
14           totPktsSent ++
15       }
16   if (event == "r" && pktType == "AGT") {
17           totPktsRecvd ++
18       }
19  }
20  END {
21          printf("\t%g",(totPktsRecvd / totPktsSent)) >> "
       PDRvarSpeed.dat"
22  }
```

Listing 7.11: Simulation Shell script to compare two protocols

```
1  #!/bin/bash
2  # ---- protocomp.sh ----
3  #run: bash protocomp.sh
4  # This script compares two protocols DSDV and AODV
5  # measures the packet delivery ratio w.r.r network load and
       speed
```

```
6
7  DIR="/home/ajit/ns-allinone-2.35/ns-2.35/indep-utils/cmu-scen-
       gen" ;# write your actual path here
8  SETDEST="$DIR/setdest/setdest"
9  CBRGEN="$DIR/cbrgen.tcl"
10 SIMULATION="./protoComp.tcl"
11 let simTime=100
12 let xCord=1000
13 let yCord=1000
14 let pause=10
15 let nodes=100
16
17   echo
18   echo "====================================="
19   echo "    Varying Network Load"
20   echo "====================================="
21   echo "Generating Scenario..............."
22   $SETDEST -v 2 -n $nodes -s 1 -m 1 -M 17 -t $simTime -P 1 -p
       $pause -x $xCord -y $yCord > ./scen
23   echo "# Packet Delivery Ratio with varying Load" > PDRvarLoad.
       dat
24    echo "# Conn DSDV    AODV" >> PDRvarLoad.dat
25    echo "# ===================" >> PDRvarLoad.dat
26    for  Conn in  5 10 15 20 25 30 35 40 45 50
27    do
28       echo -n "  $Conn" >> PDRvarLoad.dat
29       for  RP in DSDV AODV
30       do
31        echo "Generating Connection for Conn=$Conn RP=$RP..."
32        ns $CBRGEN -type cbr -nn $nodes -seed 986785 -mc $Conn   -
       rate 6.0  > ./conn$Conn-cbr
33         echo "Simulating for Conn=$Conn RP=$RP..."
34         ns $SIMULATION -rp $RP -cp ./conn$Conn-cbr -sc ./scen
       -nn $nodes -time $simTime -x $xCord -y $yCord  -tr ./CBR-
       RP_$RP-Conn_$Conn.tr
35         echo "Extracting data for Conn=$Conn RP=$RP..."
36         awk -f pdrLoad.awk < ./CBR-RP_$RP-Conn_$Conn.tr
37         rm ./CBR-RP_$RP-Conn_$Conn.tr
38         rm ./conn$Conn-cbr
39       done
40       echo >> PDRvarLoad.dat
41    done
42    rm ./scen
43   echo
44   echo "==============================================="
45   echo "    Varying Mobility speed of nodes"
46   echo "==============================================="
47   echo "Generating Connection..............."
48    ns $CBRGEN -type cbr -nn $nodes -seed 986785 -mc 25 -rate 6.0
        > ./conn-cbr
49    echo "# Packet Delivery Ratio with varying  Speed" >
       PDRvarSpeed.dat
50     echo "# Speed  DSDV     AODV" >> PDRvarSpeed.dat
51     echo "# =====================" >> PDRvarSpeed.dat
52   echo
53    for speed in  1 5 10 15 20 25 30 35
```

```
54    do
55     echo -n "  $speed" >> PDRvarSpeed.dat
56      for RP in DSDV AODV
57      do
58      echo "Generating Scenario for Speed=$Speed RP=$RP..."
59      echo
60      $SETDEST -v 2 -n $nodes -s 1 -m 1 -M $speed -t $simTime -P
      1 -p $pause -x $xCord -y $yCord >./scen$speed.scn
61      echo "Simulating for Speed=$speed RP=$RP..."
62      ns $SIMULATION -rp $RP -cp ./conn-cbr -sc ./scen$speed.scn
      -nn $nodes -time $simTime -x $xCord -y $yCord -tr ./CBR-
      RP_$RP-Scen_$speed.tr
63      echo "Extracting data for Speed=$speed RP=$RP..."
64      awk -f pdrSpeed.awk < ./CBR-RP_$RP-Scen_$speed.tr
65      rm ./CBR-RP_$RP-Scen_$speed.tr
66      rm ./scen$speed.scn
67      done
68      echo >> PDRvarSpeed.dat
69     done
70     rm ./conn-cbr
```

Once these scripts are written, execute the shell script in Listing 7.11. When the simulation is finished, it produces two output files. The output file PDRvarLoad.dat stores the output data related to varying network load simulation and PDRvarSpeed.dat stores data related to varying mobility speed. The content of these two files are given below.

Output data of PDRvarLoad.dat

# Conn	DSDV	AODV
5	0.625843	0.993236
10	0.511444	0.982301
15	0.588937	0.899612
20	0.624569	0.854936
25	0.714513	0.842807
30	0.595141	0.768029
35	0.558866	0.688124
40	0.549057	0.642765
45	0.466972	0.512713
50	0.477649	0.524013

Output data of PDRvarSpeed.dat

# Speed	DSDV	AODV
1	0.611742	0.793354
5	0.722769	0.894853
10	0.549287	0.746984
15	0.499198	0.687837
20	0.433735	0.84893
25	0.44064	0.689206
30	0.271982	0.872583
35	0.362167	0.868471

Data produced in these files may be plotted further to visualize the network performance. The following Gnuplot scripts in Listings 7.12 and 7.13 may be used to plot data. When these scripts are executed, it produces plots as given in Figure 7.13 and Figure 7.14.

Listing 7.12: GNUPlot script to plot packet delivery ratio with varying network load

```
1  #============== pdrLoad.p ==============
2  set autoscale
3  unset log
4  unset label
5  set xtic auto
6  set ytic auto
7  set xlabel "Network Load (Connections)"
8  set ylabel "Packet Delivery Ratio"
9  set grid
10 plot "PDRvarLoad.dat" using 1:2 title 'DSDV' with lp lw 2, \
11 "PDRvarLoad.dat" using 1:3  title 'AODV' with lp lw 2
```

Listing 7.13: GNUPlot script to plot packet delivery ratio with varying mobility speed

```
1  #============== pdrLoad.p ==============
2  set autoscale
3  unset log
4  unset label
5  set xtic auto
6  set ytic auto
7  set xlabel "Node Mobility (m/s)"
8  set ylabel "Packet Delivery Ratio"
9  set grid
10 plot "PDRvarSpeed.dat" using 1:2 title 'DSDV' with lp lw 2, \
11 "PDRvarSpeed.dat" using 1:3  title 'AODV' with lp lw 2
```

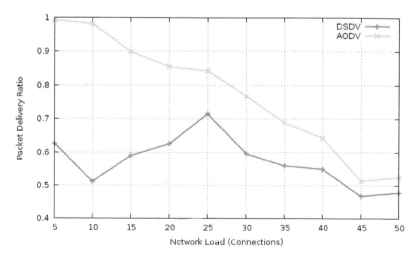

Figure 7.13: Packet delivery ratio with varying network load

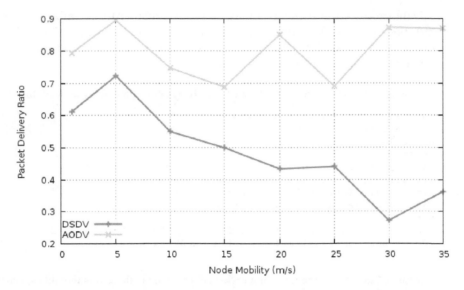

Figure 7.14: Packet delivery ratio with varying node mobility speed

Bibliography

[1] NS Mannual (http://www.isi.edu/nsnam/ns/ns-documentation.html)

[2] Marc Greies, Tutorial for the network simulator NS: http://www.isi.edu/nsnam/ns/tutorial/

[3] Altman and Jimenez, NS for beginners: http://www-sop.inria.fr/members/Eitan.Altman/COURS-NS/n3.pdf

[4] Jae Chung and Mark Claypool, NS by Example http://nile.wpi.edu/NS/Advanced Reading

Index